NOTES ON

RUBIK'S MAGIC CUBE

by

DAVID SINGMASTER

Lecturer in Mathematical Sciences and Computing
Polytechnic of the South Bank
London, England

ENSLOW PUBLISHERS
Bloy Street and Ramsey Avenue
Box 777
Hillside, New Jersey 07205

First American publication, 1981
Copyright ©1981 by David Singmaster
© 1979, 1980 by David Singmaster

Except for corrections and additional bibliography,
    this edition is identical to the British fifth edition.

Library of Congress Cataloging in Publication Data:

Singmaster, David.
    Notes on Rubik's Magic Cube.

    Includes index.
    1. Cube. 2. Polyhedra—Models. 3. Groups,
    Theory of. I. Title.
QA491.S58   1981   793.7'4   80-27751
ISBN 0-89490-043-9 paperback   AACR1
ISBN 0-89490-057-9 hardcover

Printed in the United States of America

10 9 8 7 6 5 4 3 2 1

CONTENTS.

FOREWORD

These Notes are a bit like Topsy - they have 'just growed'.
There have been so many additions for this edition that I feel it
necessary to provide a new and more substantial introduction.

The book has been retitled since the cube is now being sold both as
the Magic Cube and Rubik's Cube. This edition is twice the length of the
last version of the book and includes a table of contents, the latest
news and results, a new notation and diagrams for one face processes,
catalogues of useful processes and pretty patterns, many theoretical additions (including the discovery of $PGL(2,5)$), a bibliography, a detailed
index and a detailed step-by-step solution, found on pages 62-64.

The material of the last version has been unchanged except for
minor corrections and the liberal insertion of cross references where
there was space to add them. These are sometimes abbreviated to just
page numbers, e.g. (See pp 12,31) or (pp 12,31) or even (12,31). I have
also inserted a few names of processes or patterns when the names have
been bestowed after the original description, e.g. (4-flip; 31).
There is not always room to give such references, so the reader should
remember that a particular topic may be discussed again later and better.
The accretion of addenda to the first four editions has been retained
as the accretion of new results is felt to reflect the way in which
cubism has developed.

The basic material of these Notes is designed to give you, the
reader, a basic understanding of the Magic Cube. An algorithm for
restoring it to START is developed in section 6 and improved in section
9. A detailed step by step version of it is given separately. With
practice, this method takes less than 200 moves and less than 5 minutes.
This is quite an accomplishment, but then what? To go beyond, one must
have a good notation and some elementary notions of group theory. Section
3 introduces my notation which has been widely accepted and should be
readily understandable even to those with no mathematical background.
(Indeed, those with mathematical background often adopt much more complex
notations.) Section 4 introduces the basic concepts of permutations and
group theory, using the symmetries of the square as a standard example.
A lot of basic material is quickly covered here and the reader is advised
to skim through this section and return to it as necessary. A great
deal more group theory develops later, but always as a natural outgrowth
of playing with the cube. If the cube fascinates you as much as it does
me, you will assimilate an enormous amount of group theory and greatly
develop your spatial abilities. It is reported that Rubik actually
invented the cube to develop students' three dimensional abilities.
The cube is probably the most educational toy ever invented!

There are many exercises and problems, especially in sections 3 to
7. The reader is advised to stop and try them. Some solutions are in
section 8, but there are often better answers given later and there are
many problems which can be improved or which have not yet been examined
I hope that this presentation is accessible to those with no mathematical
background and that it will lead them up to the frontiers of the subject.
Though elementary, some of the ideas require practice and hard work to
master and the reader is advised to use lots of paper for writing out
and for drawing and to constantly have a cube in hand.

Section 1 is an earlier introduction; section 2 is generalities, including how to dismantle the cube. Sections 3 and 4 have been described above. Section 5 introduces the reader to the simpler subgroups of the cube. If your cube is in chaos, get someone to do it for you, or take it apart, or follow the step by step solution, so that you can do this section. These groups are aesthetically pleasing and are small enough that you can easily keep from getting lost. In some of them, you can't even get lost. These explorations should lead you to find sufficient basic processes that you could create a general algorithm.

Section 6 explains and later shows just which patterns of the cube are achievable. It shows how two basic processes can be used to construct an algorithm. This algorithm is greatly improved in section 9. Section 7 introduces a number of further subgroups and problems. Section 8 gives answers to the problems and some comments on the more open-ended problems.

Supplementary section 9 gives many improved processes, including the monoflip and monotwist, leading to an reasonably efficient algorithm. Section 10 presents many new results and extensions. The addenda to the first four editions extend the previous results, giving improvements and many new pretty patterns, discussing the Magic Domino, etc.

Addendum Number Five is nearly as large as all the previous material. It includes the latest anecdotes ('cubist's thumb'), discussion of Thistlethwaite's latest algorithm (at most 52 moves) and other algorithms, a new notation and diagrams for processes that only affect the U face, a small catalogue of useful processes (systematically arranged) and comments on new partial processes (Monoswop, Rubik's Duotwist, Thistlethwaite's Tritwist), a systematic catalogue of pretty patterns and an number of analyses of subgroups (square group, two generator group), among a number of other theoretical results and problems. Some rather advanced topics of group theory turn up on the cube, including PGL(2,5) in its degree 6 representation and wreath products. A detailed index (5 pages), a bibliography (3 pages), and the step by step solution (4 pages) are included.

I should like to thank the many correspondents who have provided corrections and additions. I am happy to receive further correspondence, though it is becoming overwhelming! In particular, I am thinking of compiling/editing a more theoretical book on the cube, perhaps in one of the Lecture Notes series and I would be happy to hear from potential contributors (but please try to keep in agreement with my notations!). My address is: David Singmaster
Department of Mathematical Sciences and Computing
Polytechnic of the South Bank
London SE1 OAA

My thanks to Cornelius Caesar, Jerrold Grossman, Nicholas Hammond, Michael Holroyd, 3-D Jackson, Kathleen Ollerenshaw, Itsuo Sakane and Morwen Thistelthwaite for finding errata, which now have been corrected.

Finally, I would like to dedicate this to my wife Deborah, who correctly recognised the cube as an enemy on first sight two years ago and who has wisely refused to touch it, but who has nonetheless gallantly proofread the text.

David Singmaster

NOTES ON RUBIK'S MAGIC CUBE

1. INTRODUCTION.

These notes are intended to be read in parts. Numerous exercises
and problems are given. The reader is invited to stop after reading a
problem and try to solve it. If you get stuck or are impatient, try
reading the next few sentences. The answer is rarely given immediately—
instead there may be development of the ideas involved or hints or the
solution is given in section 8.

Many of the later problems are not yet completely solved and many
undoubtedly have better solutions than given here. I would be grateful
for comments.

A Supplement of better results, an improved algorithm and further
problems has been added as sections 9 and 10 to this revised version.

2. GENERALITIES.

The action of the cube is that each face (of nine unit cubes) can be
turned about its central point. WARNING!! It only takes about 4 random
moves for the cube to become thoroughly confused. (But see page 38.)

The Basic Mathematical Problem is to restore the cube from any random
pattern back to the original pattern in which each face is of a single
colour. Masochists with a mathematical background may wish to start
solving this problem at this point. WARNING. Two weeks seems to be
the average solution time if you work very hard at it and are clever or
have some hints or know some group theory or have exceptional three-
dimensional ability. Some group theory, a notation and some introductory
hints are given below, before a solution is outlined. I recommend even
masochists to read the section on notation before beginning.

For the mechanically minded, there is the Basic Mechanical Problem of
how the cube is constructed. The internal structure is remarkably simple
and ingenious. Problem. Can you devise a mechanism for the cube? Very
few people can. The cube was invented by an ingenious Hungarian sculptor,
architect and designer, Professor Ernö Rubik, of the School for Commercial
Artists in Budapest. Though he did not work out the formal mathematics of
the basic mathematical problem, he did find a workable algorithm. (37,38)

The Hungarian instructions are fairly general and not necessary though
I have prepared a literal translation of them (my thanks to J. Dénes and A.
Kaposi for this). However, the instructions do state that the cube can be
washed in lukewarm slightly soapy water. I find that turning the cube under
a running tap works well. This removes the plastic dust which accumulates
internally and which may make turning rather stiff. The plastic is somewhat self lubricating but turning may be quite stiff at first. A lubricant
can be applied when the cube is disassembled but it is not clear what should
be used. Graphite powder (which would be ideal) or a liquid will come out
on the hands. I tried talcum powder once but it didn't seem to make any

difference. Soap, candlewax or silicone grease might work. (pp 31,37,38)

The unit cubes are not solid but are hollow plastic mouldings. One internal face has a cover plate pressed in and perhaps glued. A loose plate will give great trouble in turning. It can be pressed in and reglued (if necessary) with any ordinary glue when the cube is disassembled. A plate which is uneven or a ridge of glue can also be troublesome and these can be filed smooth with a nail file or simply allowed to wear down with use.

In turning a face, it helps to hold the cube firmly so the unit cubes stay aligned. If the cube gets a bit skew, pressing it against a flat surface will 'square' it up.

<u>Disassembly.</u> Consider any face of the cube. It contains 9 unit cubes: 4 corners, 4 edges and a centre. Rotate the top by 45° (i.e. an eighth of a turn or half-way to the next position). An edge piece of the top face can then be twisted upward, as shown by the arrow in Figure 1, and then it will come out. If the cube is stiff, a screwdriver may be used. Once one piece is out, the other pieces can be easily removed. The internal mechanism can then be inspected and will become clear. The elegant simplicity of the mechanism is truly remarkable and is a brilliant piece of three-dimensional design! To reassemble, return the pieces one by one. You will find it convenient to omit an edge of the middle layer and to turn the top layer as you put pieces into it. Put the omitted edge piece in last, in the configuration of Figure 1, simply by pressing it in and down. This may require considerable effort with a new cube. You may find some edges easier to insert than others. <u>WARNING.</u> It is advisable to reassemble in the starting pattern. We shall see that not all patterns are accessible from the starting pattern. <u>Problem.</u> Show this.

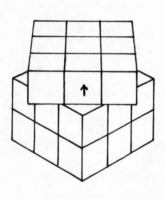

<u>FIGURE 1</u>

The face centre pieces are attached to the central spindle by a spring-loaded screw. In earlier batches of the cube, it was common for a face centre piece to come unscrewed or for its spring to get entangled with the screw, making the face too loose. If either of these happens, peel off the sticky coloured square from the face centre piece to reveal its cover plate. If the piece is off, you may be able to pop off the cover plate by pressing the screw upward into the piece or you may be able to lever it off. Otherwise, make a small neat hole in the cover plate sufficient to insert a small screwdriver or to remove the screw and spring. If you have a tangled spring, disentangle it and use a pliers to compress the top turn of the spring so it will not slip over the head of the screw again. Whichever problem you have, you can prevent repetitions by putting some glue in the screw hole of the central spindle, lubricating the head of the screw (graphite is ideal) and by turning mostly clockwise.

If you dislike the colours on your cube, you can get sticky coloured material at a stationer's. One friend has put pink dots on one colour since he had trouble distinguishing two of the colours.

Very rarely, a cube arrives with some of the pieces accidentally glued together. You can carefully separate them with a sharp knife and then smooth off the inner surfaces. (See page 60.)

3.   NOTATION.

In order to develop a notation, we need to make some first basic observations. <u>Problem.</u> What happens to the centre cubes of the faces as the faces are turned? What can you say about what happens to the edge and corner pieces?

In fact, the observations required here border on the trivial. With a little <u>careful</u> playing, you will make these observations, though you may not consciously express them. I will give the answers shortly.

Meanwhile, I will point out that the starting pattern on the cube varies from cube to cube. <u>Problem 1.</u> How many ways are there to place six colours on the six faces of a cube? (We say two colourings are the same if one can be changed to the other by a movement of the whole cube.)

The observations wanted in the first paragraph are: the centre of a face always remains in place (though it does get turned - see section 7-A); the corner pieces are moved to corner places; the edge pieces are moved to edge places. Hence we can never move a corner piece to an edge place, etc.

Thus we can always tell what the starting colour of a face was, simply by looking at its centre. Further, there can only be one form of the starting position, i.e. there is only one possible way to have each face of a single colour. We shall refer to this position as START (or GO or Square One (Cube One?)).

Conveniently, the six colours used all have different initials in English: Blue, Green, Orange, Red, White, Yellow, so we can use these initial letters for denoting the faces. However, as remarked above, the arrangement of colours varies from cube to cube. Further it is useful to have a notation which is independent of the orientation of the cube. (<u>Problem 2.</u> How many orientations does a cube have?) Consequently, it is best to develop a notation which is independent of colours and orientations. Inconveniently, the natural English words give Bottom/Back and Right/Rear problems, so I have chosen the following six names: Right, Left, Front, Back, Up, Down, which will be abbreviated by their initials.

We can now use our six letters to describe the six faces and the various pieces and positions, e.g. UR, UF, UL, UB are the four edge pieces of the U face and URF, UFL, ULB, UBR are the four corner pieces. Note that UR and RU are the same piece. We agree to list the colours or directions at a corner in clockwise order. Then URF, RFU and FUR are the same piece. We shall draw diagrams showing the F, U and R faces as in Figure 2. Generally, we will only show the colours on the F, U, R faces, but, when necessary, we can indicate colours on the unseen faces by letters around the edge, as done in Figure 2. However, Figure 2 does not completely define the pattern on the cube (<u>Problem 3.</u> Why not?) but will be adequate for our purposes.

We further use our six letters to describe the six face movements. For right-handed people, the most natural turning seems to be the following. Hold the cube in the left hand with U up and F front (i.e. towards you) so L is against the left palm. Grasp the R face with the

FIGURE 2

right hand with the thumb at D and the fingers at U (or at F and B) and turn 90° clockwise, i.e. so the upper edge turns away from you. This process, applied to the starting pattern, yields the pattern of Figure 3. (We write our letters upright even if we know they have turned.) We denote this clockwise quarter turn of the R face by R, and likewise for all the other faces. We use R^2 to denote a 180° turn and R^3 or R^{-1} to denote a 270° turn, i.e. a 90° turn anti-clockwise, and likewise for the other faces. (p 40)

We record a sequence of moves by writing them from left to right. E.g. RU means first apply R, then apply U. <u>Problem 4.</u> Draw the patterns produced by RU and by UR. Are they the same?

Before leaving this section, I will remark that in certain 'subgroups', there are other ways of denoting moves and that these may be easier to use in the subgroups. (pp 20,21,36)

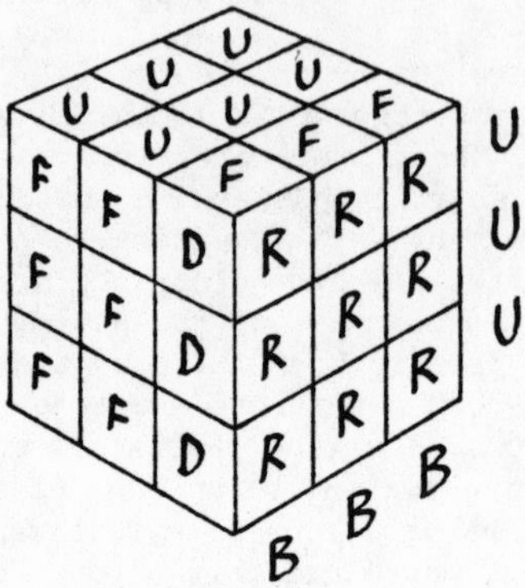

FIGURE 3

4.   SOME GROUP THEORY.

This section is quite elementary and may be skipped or skimmed at first and returned to as necessary.

Any sequence of moves rearranges the 54 (= 6 × 9) coloured faces of the unit cubes. (Actually, we may reduce this to 48 since the 6 face centres never move.) Such a rearrangement is called a <u>permutation</u>. In order to understand what patterns are possible and to phrase and study other questions, we need some terminology and some results about permutations and groups of permutations.

It is easiest to describe these ideas in a simpler context. Consider a square with corners labelled A, B, C, D as in Figure 4. A <u>symmetry</u> of

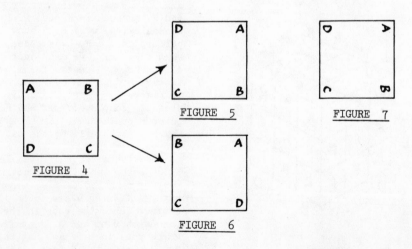

FIGURE 4

FIGURE 5

FIGURE 6

FIGURE 7

the square is a movement of the square as a rigid body which returns it to its original place, but generally with the corners in different places. Figures 5 and 6 show the effect of rotating the square clockwise by 90° and of reflecting it in the bisector parallel to AD. If we had an actual square, we could put it back where it had been, but with drawings we must imagine the drawn result as occupying the original place though it is drawn elsewhere. We conventionally always place the labels right way up, though if they were attached to the square, they would be in a different configuration. This is shown in Figure 7, which is how Figure 5 'ought' to have been drawn. We shall denote our two example symmetries of the square by R (for Rotation) and V (for Vertical reflection). <u>Problem 5.</u> Describe all the symmetries of the square.

Each symmetry of the square gives a permutation of the labels A, B, C, D. There are several ways to think of this permutation, depending on two decisions as to points of view. Firstly, we can either think of the action as 'is carried to' or 'is replaced by'. Secondly, we can either think of the permutation as acting on the labels or symbols, regardless of where they are, or as acting on the contents of positions, regardless of what symbol is presently in that position. In the latter case, we will use the labels to represent the corner positions and we can make this clearer by putting the labels outside the square as in Figure 8.

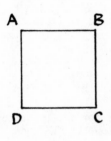

The first distinction is a duality equivalent to the duality between a permutation and its inverse. The second distinction is more significant and one point of view may be much more suitable for a given problem than the other. However, these distinctions do not become noticeable until we multiply permutations a bit later.

We can represent permutations as in the following diagrams, where the arrows represent 'is carried to' or 'is replaced by'.

FIGURE 8

	Is carried to	Is replaced by
R =	A B C D ↓ ↓ ↓ ↓ B C D A	A B C D ↓ ↓ ↓ ↓ D A B C
V =	A B C D ↓ ↓ ↓ ↓ B A D C	A B C D ↓ ↓ ↓ ↓ B A D C

The arrows and the upper rows can clearly be omitted, e.g. leaving R as B C D A. Note that the 'is replaced by' form of a permutation is the same as the 'is carried to' form of its inverse. This is the same as noting that reversing the arrows of one form gives the other. (This requires rearranging the representation somewhat - e.g. reversing arrows in the first form of R gives

$$\begin{matrix} A\ B\ C\ D \\ \uparrow\ \uparrow\ \uparrow\ \uparrow \\ B\ C\ D\ A \end{matrix}, \text{ which is } \begin{matrix} B\ C\ D\ A \\ \downarrow\ \downarrow\ \downarrow\ \downarrow \\ A\ B\ C\ D \end{matrix}, \text{ or } \begin{matrix} A\ B\ C\ D \\ \downarrow\ \downarrow\ \downarrow\ \downarrow \\ D\ A\ B\ C \end{matrix}.)$$

We write B = R(A), etc., when using 'is carried to' and D = R(A), etc., when using 'is replaced by'.

The result of applying first R and then V is called the product of R and V and we will write it RV. <u>WARNING.</u> Most books would write this as VR which is more convenient in the usual contexts since it agrees with the usual practice for multiplying or composing functions. The calculation of RV can easily be done using 'is replaced by' with permutations acting on symbols. Then R replaces A by D and V replaces D by C, so RV replaces A by C; B is replaced by A which is replaced by B, etc. So we get

$$RV = \begin{matrix} A\ B\ C\ D \\ \downarrow\ \downarrow\ \downarrow\ \downarrow \\ C\ B\ A\ D \end{matrix}, \text{ or simply C B A D, which is the reflection of the square in}$$

the BD diagonal. (The usual mode of calculation would be: (VR)(A) = V(R(A)) = V(D) = C and this is the convenience of the usual representation of the product as VR.) One can compute, with some additional difficulty, the result RV using 'is carried to' and this gives the same result as above.

The use of permutations acting on symbols means that V interchanges A and B, wherever they may be. E.g. after applying R, V is the reflection in the horizontal bisector of the square, which is the bisector parallel to AD in the turned square. If we calculate with permutations acting on positions, then V will always be reflection in the vertical bisector, interchanging the contents of the NW and NE corners (i.e. the corners where A and B originally were) and the SW and SE corners, regardless of the current contents of these positions. Then RV is
$$\begin{array}{cccc} A & B & C & D \\ \downarrow & \downarrow & \downarrow & \downarrow \\ A & D & C & B \end{array}$$
or A D C B, which is easily computed using 'is carried to'. We have that R carries A to B and V carries B to A, so RV carries A to A (i.e. it takes whatever was in the A corner to the A corner), etc. One can see that the calculation of RV using 'is replaced by' acting on positions gives the same answer.

The dichotomies between 'is carried to' and 'is replaced by' and between symbols and positions are rarely made this explicit in texts and often confuse beginners because 'is replaced by' acting on symbols, which is the usual form (corresponding to composition of functions) does not correspond to the physical symmetries in the way one would at first expect.

Problem 6. Using (a) 'is replaced by' on symbols and (b) 'is carried to' on positions, compute VR, VR^2, VR^3.

I shall generally use 'is carried to' on positions in describing permutations on the cube, since the basic moves R, U, ..., are defined in terms of physical positions. However, there are certain 'subgroups' in which one can use 'is replaced by' on symbols.

Now we wish to describe a different representation for a permutation which will be more informative. Consider any permutation P. The successive applications of P are denoted P, P^2, P^3, ... These permutations carry (or replace) a position (or symbol) A by P(A), $P^2(A)$, $P^3(A)$, ... Beginning with any convenient position (or symbol) A, this sequence must eventually repeat (i.e. cycle). Because of the one to one (or invertible) nature of a permutation, this sequence can only repeat by first returning to A at some stage, say $P^n(A) = A$ and $P^i(A) \neq A$ for $1 \le i < n$. We represent this by enclosing one complete cycle in parenthesis thus: (A, P(A), $P^2(A)$, ..., $P^{n-1}(A)$). This is called an n-cycle. E.g. for R on the square, using 'is carried to' on symbols, we get (A, B, C, D). We can graphically represent cycles as in Figures 9 and 10. Since the starting point of a cycle is arbitrary, the cycle (A, B, C, D) is the same as (B, C, D, A), etc.

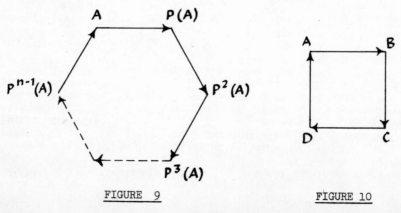

FIGURE 9   FIGURE 10

If there are any symbols (or positions) left over, we take any convenient one as the beginning of another cycle and continue till all symbols (or positions) are exhausted. E.g. for V, we get two 2-cycles (A,B)(C,D) which we draw as in Figure 10', while RV gives (A)(B,D)(C) which is drawn in Figure 11. We usually omit cycles of length one, so we write

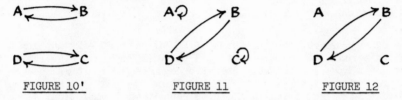

FIGURE 10'   FIGURE 11   FIGURE 12

RV simply as (B,D), as drawn in Figure 12. We say RV acts only on B and D since it leaves all the other objects fixed. <u>Problem 7.</u> Find the cycle representation of each symmetry of the square.

Consider now the moves on the cube. We consider permutations using 'is carried to' acting on positions. The cycle representation of the move R is (FR,UR,BR,DR)(URF,BRU,DRB,FRD) and the cycle representation of RU is (FR,UF,UL,UB,UR,BR,DR)(URF,RFU,FUR)(BRU,DRB,FRD,UFL,ULB,UBR,...). Here we run into a notational complexity. The second cycle of RU involves three triples which are the same corner of the cube but in different orientations. That is, RU twists URF by 1/3 turn clockwise (as viewed from outside the cube). We shall simply write this as (URF)$_+$ where the subscript + denotes a clockwise twist. Equivalently, + indicates that each label is carried to the next one in the sequence URF. We shall call this a <u>twisted</u> 1-cycle. The third cycle of RU should have 15 entries but they comprise three repetitions of the first 5 corners, with each repetition obtained by turning the previous ones 1/3 turn anti-clockwise. We denote this twisted 5-cycle by (BRU,DRB,FRD,UFL,ULB)$_-$. This means the fifth power of this cycle will be (BRU)$_-$(DRB)$_-$(FRD)$_-$(UFL)$_-$(ULB)$_-$. We can also have twisted cycles of edges, but since an edge piece has only two sides, + and − are the same and we will write (FR)$_+$ for (FR,RF), etc. <u>Problem 8.</u> Find the cycle representations of R^2 and UR.

We now return to general considerations of permutations. If two permutations P, Q act on disjoint sets, then clearly PQ = QP. E.g. (A,B)(C,D) = (C,D)(A,B). Thus the order of disjoint cycles is irrelevant. We say P and Q <u>commute</u> if PQ = QP. (This can happen when P and Q do not act on disjoint sets, e.g. $RR^2 = R^2R$.) We then have that $(PQ)^2 = PQPQ = PPQQ = P^2Q^2$, $(PQ)^3 = P^3Q^3$, etc.

There is a special permutation, the <u>identity</u> I, which doesn't move anything. On the square, I has the representations

```
A B C D
↓ ↓ ↓ ↓      or      (A)(B)(C)(D)   or                      ,
A B C D
```

where the last, empty, cycle representation is subject to misreading so we denote it by I. Note that IP = P = PI for any permutation P and that I acts on no objects.

For any permutation P, there is an <u>inverse</u> permutation P^{-1}, which inverts the action of P. If P(A) = B, then P^{-1}(B) = A. In our first representation of P, P^{-1} is obtained by reversing all the arrows. The cycle representation of P^{-1} is obtained by reversing each cycle of P.

E.g. for the square, R^{-1} has the representations:

```
A B C D        A B C D        A ← B
↑ ↑ ↑ ↑   or   ↓ ↓ ↓ ↓   or   ↓   ↑   or   (D,C,B,A)  or  (A,D,C,B).
B C D A        D A B C        D → C
```

---

* <u>Note.</u> FR → UR means that the F side of FR goes to the U side of UR, etc.

R^{-1} corresponds to rotating the square $90°$ anti-clockwise. For any permutation P, we have that $PP^{-1} = I = P^{-1}P$. Generally, for permutations P, Q, we have $(PQ)^{-1} = Q^{-1}P^{-1}$ (i.e. to undo something, you undo all the steps in reverse order) and this is generally not equal to $P^{-1}Q^{-1}$ unless P and Q commute. Exercise. Check by examining $(PQ)(Q^{-1}P^{-1})$, $RUU^{-1}R^{-1}$ and $RUR^{-1}U^{-1}$.

A consequence of the existence of P^{-1} is that $PQ = PR$ implies $Q = R$. (We haven't really demonstrated that a permutation has an inverse. It can be shown by a general set theoretic argument if we formalise the definition of permutation and it follows from the next paragraph, but I think that it is sufficiently clear to require no formal proof.)

Now consider any permutation P written as a product of disjoint cycles, say $P = C_1C_2C_3$. Then $P^2 = C_1^2C_2^2C_3^2$, ..., $P^m = C_1^mC_2^mC_3^m$ and $P^{-1} = C_1^{-1}C_2^{-1}C_3^{-1}$. Consider now any n-cycle $C = (A_1,A_2,...,A_n)$. Now C carries A_1 to A_2, C^2 carries A_1 to A_3, C^3 carries A_1 to A_4 and generally C^i carries A_1 to A_{1+i} and A_j to A_{j+i}. (When j+i exceeds n, we cycle the subscript back to 1. E.g. A_{n+1} is A_1, A_{n+2} is A_2, etc.) C^i may consist of more, smaller, cycles than C. E.g. on the square, $R = (A,B,C,D)$, so $R^2 = (A,C)(B,D)$, $R^3 = (A,D,C,B)$, $R^4 = (A)(B)(C)(D) = I$. Note that C^n carries each A_j to $A_{j+n} = A_j$ so that $C^n = I$. Hence we have $C^{n-1} = C^{-1}$, e.g. $R^3 = R^{-1}$ since we know $R^3R = R^4 = I$. For the cube, $(RU)^2 = RURU$ is (FR,UL,UR,DR,UF,UB,BR)(URF)_(BRU,FRD,ULB,BDR,LUF)_+. Problem 9. Write down $(RU)^3$, $(RU)^{-1}$ and $(RU)^{15}$.

A 2-cycle, like (A,B), is called a transposition or interchange. Note that $(A,B)^2 = I$; $(A,B)^i = I$ if i is even; $(A,B)^i = (A,B)$ if i is odd; $(A,B)^{-1} = (A,B)$ and $(B,A) = (A,B)$. It is not hard to see that any n-cycle can be written as a product of n-1 (not disjoint) transpositions. E.g. $(A,B,C) = (A,B)(A,C)$. Problem 10. Write (A,B,C,D) as a product of 3 transpositions. Hence any permutation is a product of transpositions, but in many different ways, e.g. $I = (A,B)(A,B) = (A,C)(A,C)$. It is a remarkable fact that the number of transpositions in these different ways is always even or always odd. Permutations are classified as even or odd according to whether they are a product of an even or an odd number of transpositions.

{For those wishing to consider this point further, let the objects being permuted be numbered 1, 2, 3, ..., n. Let $P(i) = j$ mean that the i-th object is permuted to (i.e. is carried to or is replaced by) the j-th object. Let x_1, x_2, ..., x_n be n distinct numbers and consider

$$\text{Sign}(P) = \frac{\Pi_{i<j}\ (x_i - x_j)}{\Pi_{i<j}\ (x_{P(i)} - x_{P(j)})} .$$

It is not hard to see that Sign(P) is +1 or -1 and that $\text{Sign}((A,B)P) = -\text{Sign}(P)$, whence the sign of a product of s transpositions is just $(-1)^s$. Hence Sign(P) = +1 if P is even and Sign(P) = -1 if P is odd.}

It follows that even·even = even = odd·odd and even·odd = odd = odd·even, that is, the combination of even and odd permutations is like the addition of even and odd integers. An n-cycle is even or odd depending on whether n-1 is even or odd. Problem 11. Classify the symmetries of the square as odd or even. On the cube, show that all patterns are even permutations of START, considered just on the pieces, not on their orientations. Hence one can never obtain, e.g. (UR,UL) or (URF,UFL).

The <u>order</u> of a permutation P is the least positive integer m = ord(P) such that P^m = I. (We shall only consider permutations where such an m exists.) We have that ord(P) = 1 if and only if P = I. The order of an n-cycle is n. On the cube, the order of a twisted n-cycle of edges is 2n and the order of a twisted n-cycle of corners is 3n. From the fact that P is invertible, $P^i = P^j$ implies P^{i-j} = I. We now show that m divides i-j. We write i-j = mq + r where $0 \le r < m$, i.e. we divide m into i-j to get a quotient q and a remainder r. Then r = i-j-mq and $P^r = P^{i-j}P^{-mq} = I(I)^{-q} = I$ so r is a smaller positive integer than m, with P^r = I, contrary to our definition of m, unless r is 0, i.e. m divides i-j exactly. Consequently we have $P^i = P^j$ if and only if m divides i-j and P^i = I if and only if m divides i. Thus the powers $I(= P^0)$, P, P^2, ..., P^{m-1} form a cycle of length ord(P).

If P is written as a product of disjoint cycles, say $P = C_1 C_2 C_3$, then $P^m = C_1^m C_2^m C_3^m = I$ if and only if $C_1^m = I$, $C_2^m = I$, $C_3^m = I$ which is if and only if m is a multiple of the lengths (i.e. orders) of each cycle. That is, ord(P) is the least common multiple (LCM) of the orders of the C_i's. E.g. for the square, ord(R) = 4, ord(V) = 2, while on the cube, ord(R) = 4, ord(RU) = LCM(7,3,15) = 105, i.e. $(RU)^{105}$ = I but no smaller positive integral power of RU is I. One can show that if ord(P) = m, then we have ord(P^i) is m divided by the greatest common divisor (= highest common factor) of m and i.

Consider now <u>all</u> the permutations of the four letters A,B,C,D. There are 4·3·2·1 = 24 of these. In general the total number of permutations of n things is n(n-1)(n-2)...3·2·1 and this product is denoted <u>n!</u> (pronounced n factorial). It is convenient, consistent and conventional to set 0! = 1, though we do not require this.

The set of all permutations of n objects forms a <u>group</u>, which is a set G with operations of multiplication and inversion and containing I, such that:

a)   if P, Q are in G, then so is PQ (closure under multiplication);
b)   I is in G (existence of I or closure under I);
c)   if P is in G, then so is P^{-1} (closure under inversion);
d)   if P, Q, R are in G, then P(QR) = (PQ)R (associative law);
e)   if P is in G, then PI = P = IP;

f)   if P is in G, the $PP^{-1} = I = P^{-1}P$.

Properties d, e, and f hold for any permutations. We have already noted properties e and f as general properties of I and P^{-1}. Property d is true for any functions P, Q, R since (P(QR))(A) and ((PQ)R)(A) are both, in the conventional notation, just P(Q(R(A))) {or R(Q(P(A))), in our notation}.

Consider now any subset H of a group G. This will again be a group, under the same operations, if properties a to f hold. Properties d, e, f hold automatically in H since they hold in the bigger set G. If H is a finite set, then property a implies properties b and c. (Let P be in H. Then property a implies P^2, P^3, ... are all in H. Since H is finite, P must have some finite order, say m. Thus P^m = I and $P^{m-1} = P^{-1}$ are in H.) A subset of a group which is again a group under the same operations is called a <u>subgroup</u>. The group G of all permutations on a finite set of objects is finite, so any of its subsets are finite. Hence a subset of G is a subgroup if and only if it is closed under multiplication. (The group G is commonly denoted S_n if there are n objects being permuted.)

For example, the symmetries of **our square** are a subgroup of the permutations of A, B, C, D, since the product of two symmetries is again a symmetry. Using this example, we can illustrate the two basic ways of creating and describing subgroups. Firstly, we can describe a subgroup as the set of permutations which preserve some structure or property. E.g. the symmetries of the square preserve its 'squareness'. (This can be more explicitly described as saying the letters A and C are never adjacent in representations such as R = B C D A. For the cube, we have 48 moveable unit coloured faces (ignoring the face centres) and there is a group of all permutations of these 48 objects. The set of all patterns reachable from START forms a subgroup which might be said to preserve the 'cubicity' in some sense.

Secondly, we can describe a subgroup as the smallest subgroup H containing some given set S of permutations. This smallest subgroup consists of all finite products of the given permutations (and their inverses if G is infinite). <u>Problem 12.</u> Prove this. We say H is the subgroup <u>generated</u> by S. If S = {P, Q, ...}, we write H = <P, Q, ...>. E.g. the symmetries of the square are generated by R and V. <u>Problem 13.</u> Verify this by showing that the symmetries of the square are the following:

I, R, R^2, R^3, V, RV, R^2V, R^3V. The subgroup generated by R, i.e. <R>, is {I, R, R^2, R^3}, which is the group of direct symmetries or rotations (i.e. reflections are not allowed) of the square. $<R^2>$ = {I, R^2} and <V> = {I,V}. The patterns on the cube are the subgroup generated by {R, L, F, B, U, D}.

<u>Problem 14.</u> In any group of permutations, the even permutations form a subgroup. If a group contains any odd permutations, then exactly half of its permutations are odd. E.g. the even subgroup of the symmetries of the square is {I, R^2, V, R^2V}. As another example, consider the symmetries of a regular tetrahedron. This is the same as the group of all permutations of A, B, C, D, if we allow reflections. If we permit only rigid motions in space, we get just the even permutations of A, B, C, D. However, as the square shows, the <u>direct</u> symmetries (i.e. those when reflections are not permitted or those which preserve orientation) are not always the even permutations.

A permutation of order m generates the subgroup {I, P, P^2, ..., P^{m-1}}, which is an example of a <u>cyclic</u> subgroup of <u>order</u> m. In our discussion of even permutations, we saw that the group of all permutations of a set of objects is generated by the set of transpositions (i.e. 2-cycles). For the moment, let us assume that our set of objects is {1, 2, 3, ..., n}. <u>Problem 15.</u> Show the group of all permutations on this set is generated by {(1,2), (1,3), (1,4), ..., (1,n)}. <u>Problem 16.</u> Show the subgroup of **even permutations is generated by the set af all 3-cycles** or by the set of all pairs of disjoint 2-cycles or by {(1,2,3), (1,2,4), (1,2,5), ..., (1,2,n)} or by {(1,2)(3,4), (1,2)(3,5), ..., (1,2)(3,n), (1,3)(2,4)}. (Assume n ≥ 6.)

5. SOME EXPLORATIONS.

In order to understand the behaviour of the cube, it helps to first study some small or simple subgroups and to discover some processes which move only a few pieces. I give a few examples below, though you may want to start exploring on your own first.

A. The Slice Group.

(This group and its name were described to me by John Conway.)

Holding the cube in the standard position, as used on page 3 in describing the move R, imagine turning both the R face and the L face away from you. This is equivalent to turning the central 'slice' toward you and would be denoted by RL^{-1} (= $L^{-1}R$) in our notation. Consider the subgroup generated by these slice moves. It is convenient when working in this group to let R denote the just described slice motion (i.e. RL^{-1} in our original notation). Notice then that $R = L^{-1}$ in this notation, so this group is generated by the three slice moves R, U, F.

This group is not too large and contains some elegant patterns. It is recommended to novices who want to play with the cube without getting too lost. A little examination shows that each face will always display a pattern of the type shown in Figure 13, where a, b, c, d are four colours (not necessarily distinct). Further, the face opposite to this will have the same pattern with each colour replaced by its opposite. (The opposite colours are R-L, F-B, U-D.) We shall denote the colour opposite to a by a', e.g. R' = L. With some systematic playing you should be able to obtain patterns with all six faces having a = b = c ('spot' or 'box' or 'measles' face) or with all six faces having a = d = b' = c' ('X' or 'cross' face) or with four spot faces and two solid faces (i.e. a = b = c = d) or with four faces having a' = b = c = d ('+' or 'plus' or 'Greek cross' face) and two X faces. However, not all patterns which look like Figure 13 can be obtained.

a	b	a
c	d	c
a	b	a

FIGURE 13

The full structure of this group is not too complicated. You may be able to already determine the subgroup where a = b = c on all faces. To get further, it seems best to observe that the moves already defined are not the best to work with. Hint. Contemplate the corners.

B. The Slice-squared Group.

Consider the subgroup generated by the squares of the slices, i.e. $\langle R^2, U^2, F^2 \rangle$ in the notation of subsection A. Hint. Compare R^2U^2 and U^2R^2.

C. Two Squares Group.

Consider the subgroup generated by the squares of two adjacent faces, e.g. $\langle F^2, R^2 \rangle$ in our usual notation.

D. The Antislice Group.

An antislice is a movement such as RL (= LR) in our original notation. This corresponds to turning the R face away from you while turning the L face toward you. If we denote this antislice by R, we have R = L in this group. The square of an antislice is the same as the square of the corresponding slice. The structure of this group is largely unknown but it contains some pretty patterns. With considerable playing, you can obtain patterns with four 'diagonal' faces (Figure 14) and two solid faces, with four 'Z' faces (Figure 15) and two solid faces, with six '2L' faces (Figure 16). with four + faces and two solid faces and with four diagonal faces and two + faces.

a	a'	a'
a'	a	a'
a'	a'	a

FIGURE 14

a	a	a'
a'	a	a'
a'	a	a

FIGURE 15

a	a	a
a'	b	a
a'	a'	a'

FIGURE 16

E.   The Two Generator Group.

Consider the subgroup generated by two adjacent moves, e.g.
<F, R> in the usual notation.

F.   Some Simple Processes.

If you have made some systematic explorations of the above groups,
you will have found processes (i.e. sequences of moves) which interchange
two pairs of edges, leaving everything else fixed, and which interchange
two pairs of corners, leaving all other pieces (or just all other edges)
fixed. If you haven't yet found such processes, look back again at the
Two Squares Group and the Two Generator Group. You may have found processes
which give just a 3-cycle on edges or on corners. You cannot find processes
to interchange a single pair of edges or of corners (see Problem 11). It is
possible to find a process which interchanges one pair of corners and one
pair of edges, though I don't know any one which is simple. (pp 35,36,45)

Once you have found these processes, can you now see that you can
interchange <u>any</u> two pairs of edges and <u>any</u> two pairs of corners? (Here
and elsewhere, reference to two pairs means two disjoint pairs.) If you
have found 3-cycles, can you see that you can 3-cycle any triple of edges
and any triple of corners? Further can you see that you can flip any pair
of edge pieces and twist any pair of corners in opposite directions?

6.   THE BASIC MATHEMATICAL PROBLEM.

The problem is to restore the cube from any random pattern back to
its original position, START, with each face having just a single colour.
Inversely, this is also the problem of getting from START to any given
pattern and hence is also the problem of getting between any two patterns.

To solve this problem, we need to proceed in two directions. First,
by examining the cube and its group, we discover which patterns are possible
and, second, we show that we can achieve all possible patterns. In fact,
all the necessary processes have been discovered in section 5-F. We now
concentrate on the question of describing the possible patterns. If you
are ambitious, you may try to do this before going on to the next paragraph.
Even if you don't want to think exactly, perhaps you would like to make
a guess as to how many patterns are possible. <u>Hint.</u> Simple combinatorial
considerations will get you fairly close to the right number.

The group of all possible permutations is as follows. The 8 corner
pieces can be permuted among themselves in any way, giving 8! ways, and
the 12 edge pieces can be permuted among themselves in any of 12! ways,
except that the total permutation of corners and edges must be even.
Further, independently of the movement of pieces, we can flip the orientations
of any two edge pieces and we can twist any two corner cubes in opposite
directions. This means that we can orientate all but one of the edges
or corners as we please, the orientation of the last one being forced.
This means we have a total of

$$N = \frac{8! \; 12!}{2} \cdot \frac{3^8}{3} \cdot \frac{2^{12}}{2} = 43252 \; 00327 \; 44898 \; 56000 \approx 4.3 \times 10^{19}$$

$$= 2^{27} 3^{14} 5^3 7^2 11$$

different patterns. A computer might count one pattern per microsecond,
so it would take about 1.4 million years to count through all these patterns!
The numerator of N is the total number of ways you can reassemble the cube
and is the number you might have obtained in the previous paragraph. The
denominator of 12 means that there are 12 distinct orbits of constructible
patterns. (An orbit is the set of all patterns reachable from a given
pattern by application of our group. You cannot get out of one orbit into
another by use of the group of motions.) Initially, one might have expected
that all constructible patterns were achievable from START, but we see that
the group of possible patterns is only a twelfth as large as we might have

expected. This also means that if we reassemble the cube at random, there is only a 1/12 chance of being able to get back to START. (This justifies the warning given in the middle of page 2.) (This description of the possible positions was first given to me by John Conway who attributes it to Anne Scott.)

Knowing the possible patterns, we can now see that we can get all of them by using the simple moves described in section 5-F. (Can you see this?) It suffices to see that we can use our simple processes to produce any even permutation of corners, any even permutation of edges, any flip of two edges and any twist of two corners in opposite directions. The observant reader will notice that a pattern with an odd permutation of corners and an odd permutation of edges is an even permutation overall and hence is possible. However, we pass from the odd, odd case to the even, even case by any single move, so we need only consider the even, even case.

First, let us consider the edges. It suffices to show that we have all pairs of 2-cycles or all 3-cycles since these generate the even permutations (see Problem 16). You have, I hope, already discovered such processes which leave everything else fixed. A little thought and playing will show that just one of these processes is sufficient to generate all of them in the following way. Any four (or three) edge pieces can be moved to any four (or three) edge positions by a few moves. Further, the last piece can be in either of its two orientations. After doing this, we can apply our basic process on the four (or three) positions and then invert the previous moves to move the pieces back to the original positions. This accomplishes a pair of 2-cycles (or a 3-cycle) on the original positions. (In group theory, this process is called <u>conjugation</u>. The general expression for P conjugated by Q is QPQ^{-1}. The conjugate of a permutation is a permutation with the same cycle structure. Basically a conjugate does the same thing but on different objects.) (See also pp 57-59.)

My own favourite edge process is $P_1 = P_1(F,R) = (F^2R^2)^3 = (R^2F^2)^3$ which is two 2-cycles: (FU,FD)(RU,RD). Note that $P_1^2 = I$, i.e. $P_1^{-1} = P_1$, and $P_1(F,R) = P_1(R,F)$. This process is indicated in Figures 17 and 18, the latter being a schematic version of the first.

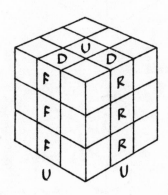

FIGURE 17

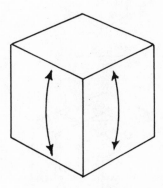

FIGURE 18

<u>Problem 17.</u> Using P_1 or otherwise, obtain the following.

A)   i)   (UF,UB)(UR,UL)
       ii)   (UF,UR)(UB,UL)
       iii)   (UF,UR,UB)
       iv)   (UF,UR,FR)
       v)   (UF,RU,FR)
B)   (FR,BR)(FL,BL)
C) $(UF)_+(DF)_+$

From Problem 17-C and the previous argument, we deduce that any flip of two edges can be achieved.

Before continuing on to corners, let us examine the process of conjugation a bit further. We will illustrate by considering Problem 17-A-i in detail. Applying FB^{-1} to START produces Figure 19, then U gives Figure 20. We are now in a position to apply $P_1(R,U) = (R^2U^2)^3$ and this will yield Figure 21. Then $U^{-1}BF^{-1}$ gives Figure 22 which is the desired pattern. Thus the solution to the problem is $FB^{-1}U(R^2U^2)^3U^{-1}BF^{-1} = QPQ^{-1}$, where $P = P_1(R,U) = (R^2U^2)^3$ and $Q = FB^{-1}U$. The process of conjugation plays a major role in the theory of groups and its obvious utility in our problem shows how useful it is in general. Using conjugation, the arguments on the previous page and the result of Problem 17C, we have that we can accomplish any possible edge manipulations. (Sections 9-A and 9-B of the Supplement describe several other edge processes.)

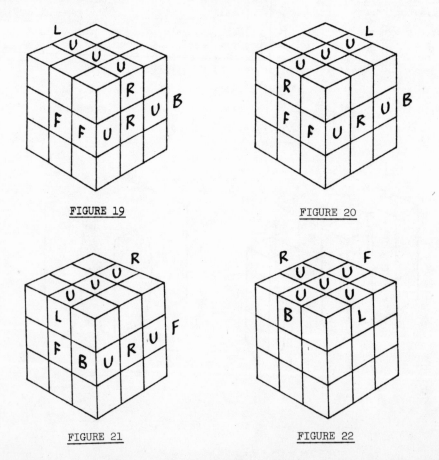

FIGURE 19

FIGURE 20

FIGURE 21

FIGURE 22

So now we can consider the corners. It suffices to show that we have all the 3-cycles or all pairs of 2-cycles. One of these is sufficient as we can move any four (or three) corner pieces into any four (or three) corner positions. Further we must see that we can twist any pair of corners in opposite directions and it will suffice to see that we can twist some one pair. Conceptually, we have the same kind of solution as for the edges. The following shows that we can carry out these processes, though the processes are not very efficient.

My processes are based on the following.

$$P_2 = P_2(F,R) = FRF^{-1}R^{-1} = (FLU,FUR)_+(FRD,DRB)_-(FU,FR,DR)$$
$$P_3 = P_2^2 = (FLU)_+(FUR)_+(FRD)_-(DRB)_-(FU,DR,FR)$$
$$P_4 = P_2^3 = (FLU,URF)(FRD,BDR)$$

These are shown in detail in Figures 23, 24, 25 and schematically in Figures 26, 27, 28.

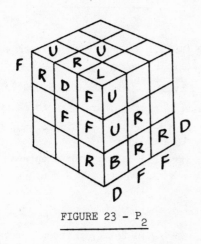

FIGURE 23 - P$_2$

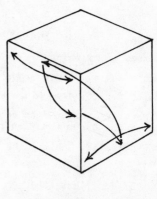

FIGURE 26 - P$_2$

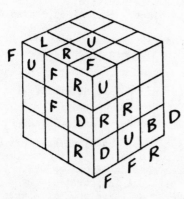

FIGURE 24 - P$_3$

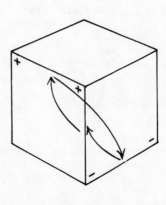

FIGURE 27 - P$_3$

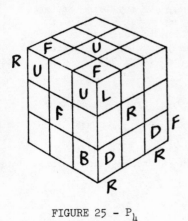

FIGURE 25 - P_4

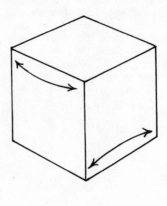

FIGURE 28 - P_4

Note that $P_2^6 = I$ and that $P_2(F,R)^{-1} = P_2(R,F)$. By combining P_2 with edge processes or by using P_4, we can move corners as we please. By combining two P_3's and some edge processes, we can twist two adjacent corners in opposite directions. Since edge processes, using P_1, are easy to carry out, I originally preferred to get all the corners correct by using P_2 and P_3, ignoring the edges and then put the edges in place using P_1.

Problem 18. Obtain the following, ignoring edges. What happens to the edges?
A) (FLU,RUB,RFU)
B) (FLU,FUR)$_+$(BRU,LBU)$_-$
C) (FLU)$_+$(FDL)$_-$
D) (FLU)$_-$(FUR)$_-$(RUB)$_-$
(Sections 9-B and 9-C of the Supplement give better methods for these results, mostly leaving the edges fixed! Section 9-D gives a much better general algorithm than described below.)

Now I summarize my original process for restoring the cube to START. This is not too efficient, but shows that it is possible.
1) Get all the bottom corners into place. (This requires no particular process.)
2) If the top corners are in an odd permutation, apply U.
* 3) Apply various P_2's to get all top corners in the correct positions.
4) Apply various P_3's to twist top corners into correct orientations.
5) Apply various P_1's to put edges into the right places with the right orientations.

I find that this algorithm takes little memory, since P_1 and P_2 are simple processes, but sometimes the conjugation operations are sufficiently complicated that I have to jot them down. I have counted the maximal number of moves that this process requires and find it to be 277. (This is a rather tedious calculation and may be a bit inaccurate.) By a move, I mean any turn or its square or its inverse as these are all single moves of the hand. Problem 19. Find a process with a smaller maximal number of moves. In my process, the lengthiest single stage is that of Problem 18-D,

* Note. The phrase 'various P_2's' implies various conjugates of P_2, etc.

where my solution requires 26 moves. <u>Problem 20.</u> Can you do Problem 18-D in less than 26 moves?

In group theory, a product $PQP^{-1}Q^{-1}$ is called a <u>commutator</u>. We have $PQP^{-1}Q^{-1} = I$ if and only if $PQ = QP$, so the commutator is an indication of whether P and Q commute. Thus our process P_2 is a commutator in our group.

I have found that different people have quite different strategies for restoring the cube to START. Several people get all the edges in place first, then apply P_4's or other combinations of commutators to get the corners in place. The process $P_2(F,R^{-1}) = FR^{-1}F^{-1}R =$ (FLU,RUB)_(FUR,FRD)_+(FU,FR,UR) is often used instead of or in combination with P_2. (R. Penrose calls $P_2(F,R)$ a Z-commutator and $P_2(F,R^{-1})$ a Y-commutator because of the arrangement of the affected corners.) Many people get the whole bottom layer correct, then the middle layer, then the top layer. Section 9-D describes such a process.

It remains for us to verify the statement that we can only flip edges in pairs and can only twist corners as we have stated. This is a somewhat tedious process but I hope the following argument is clear enough to convince you. The point is that we now consider our corner and edge positions in some definite fixed orientations, so that we can examine how a turn affects orientations. We write down the 12 edge pairs and the 8 corner triples in some arbitrary sequence such as:

UF, UL, UB, UR, ... , UFL, ULB, UBR, URF, ...

This particular sequence makes the examination of the move U easy. Applying U will permute the pieces but will not change any of the orientations of pairs of triples. However, we cannot expect that the orientations will always be so convenient. Let us see what happens if the orientation of one corner is changed, e.g. if ULB is replaced by LBU. After this replacement, the action of U on U corners is:

$$\begin{array}{cccc} UFL & LBU & UBR & URF \\ \downarrow & \downarrow & \downarrow & \downarrow \\ ULB & BRU & URF & UFL \end{array}$$

We see that the image of UFL is not LBU but is LBU shifted forward one place and the image of LBU is not UBR but is UBR shifted back one place. If we denote these shifts by +1 and -1, we have that the sum of the shifts is 0. Technically, we must consider these shifts (mod 3), i.e. 3 shifts of +1 brings us back to 0 so -1 = +2, etc. These shifts correspond to rotations of the corner piece by 120° about its main diagonal.

Now since the sum of orientation shifts is 0 for any one change of orientation, it must be 0 for any changes of orientations, i.e. it will be 0 for any orientations of the corner triples. Symmetrically, this holds for any of our six basic turns and so it is a property preserved by all members of our group. Since START is a pattern in which the sum of orientation shifts was 0, any possible pattern must have a 0 sum of shifts. In particular, if we reach a pattern in which each corner is in its correct place, we can readily see the orientation shifts as the amounts the corners are twisted and so the total twist of corners must add to 0. This total is always taken (mod 3) so that the pattern of Problem 18-D, with three twists of -1, is possible.

A similar argument shows that the sum of the edge flips must be 0 (mod 2) i.e. the number of edge flips is always even.

(See page 32 for further discussion of ways to assign orientations.)

7. FURTHER PROBLEMS.

A. The Supergroup.

Consider the centres of the faces. Though these never move, they are turned. The total turn of R in a process (measured in units of quarter turns or 90°) is just the sum of the exponents of R throughout the process. This sum must be considered (mod 4) since 4 turns is the same as none. E.g. for $P_1 = (F^2R^2)^3$, R has turned 6 units, which is the same as 2 units.

How does considering the centre turns affect the group? One can consider replacing the colour patterns on faces at START by some pictures so that the centre squares must also be restored to their correct orientations for correct pictures.

B. The Three Generator Groups.

What are the subgroups generated by three of the basic moves? There are two types, e.g. <F, U, R> and <F, U, B>. In both cases, some pieces are never moved. What if we take the squares of three basic moves?

C. The Four Generator Groups.

There are again two types, e.g. <F, U, R, D> and <F, L, B, R>. The first never moves LB. The second can never flip two edge pieces. Hence neither is the whole group. What if we take the squares of these basic moves?

D. The Five Generator Group.

R. Penrose has shown that one generator can be ignored and we still get the whole group. I have been told that he has a 28 move process to produce the effect of one turn, using only the 5 other turns.

E. The Square Group.

What is the subgroup generated by $\{R^2, L^2, U^2, D^2, F^2, B^2\}$?

F. The Maximal Order Problem.

What is the maximal order of a permutation in the group?

G. The Commutator Subgroup.

A commutator is a product of the form $PQP^{-1}Q^{-1}$. What is the subgroup generated by all the commutators?

H. Pretty Patterns.

There are undoubtedly many more pretty patterns than have been seen so far.

I. Super Problems.

One can consider almost all of the various problems discussed so far within the supergroup of subsection A rather than within the original group.

(A number of further results on these problems and some further problems are given in section 10.)

8. SOME ANSWERS AND COMMENTS.

Problem 1. (page 3) 30
Problem 2. (page 3) 24 (or 48 if reflections are permitted)
Problem 3. (page 3) As will be seen in section 6, we can indeed permute and flip the unseen edge pieces, LB, LD, BD in a number of ways (to be precise, in 12 patterns), but the unseen corner LBD will be fixed.

Problem 4. (page 4) As seen from the drawings below, they are not the same.

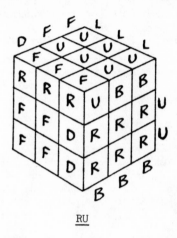

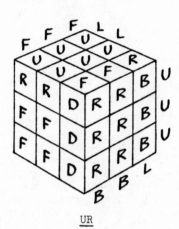

RU   UR

Problem 5. (page 5) There are 8 of them: rotations through 0°, 90°, 180° and 270° and reflections in the vertical and horizontal bisectors and the two diagonals, which we denote by I, R, R^2, R^3, V, H, D_1, D_2.

Problem 6. (page 6)

a)
$$VR \quad \begin{matrix} A & B & C & D \\ \downarrow & \downarrow & \downarrow & \downarrow \\ A & D & C & B \end{matrix} \qquad VR^2 \quad \begin{matrix} A & B & C & D \\ \downarrow & \downarrow & \downarrow & \downarrow \\ D & C & B & A \end{matrix} \qquad VR^3 \quad \begin{matrix} A & B & C & D \\ \downarrow & \downarrow & \downarrow & \downarrow \\ C & B & A & D \end{matrix}$$

b)
$$VR \quad \begin{matrix} A & B & C & D \\ \downarrow & \downarrow & \downarrow & \downarrow \\ C & B & A & D \end{matrix} \qquad VR^2 \quad \begin{matrix} A & B & C & D \\ \downarrow & \downarrow & \downarrow & \downarrow \\ D & C & B & A \end{matrix} \qquad VR^3 \quad \begin{matrix} A & B & C & D \\ \downarrow & \downarrow & \downarrow & \downarrow \\ A & D & C & B \end{matrix}$$

Changing between a) and b) is the same as reversing the order of the product.

Problem 7. (page 7) Using 'is carried to' on positions,
I = I, R = (A,B,C,D), R^2 = (A,C)(B,D), R^3 = (A,D,C,B),
V = (A,B)(C,D), H = (A,D)(B,C), D_1 = (A,C), D_2 = (B,D).

Problem 8. (page 7) R^2 = (FR,BR)(UR,DR)(URF,DRB)(BRU,FRD)
UR = $\overline{(FLU,LBU,DRB,FRD,URF)}_-(UBR)_+(FU,LU,BU,RB,RD,RF,RU)$

Problem 9. (page 8)

$(RU)^3$ = (FR,UB,DR,UL,BR,UF,UR)(BRU,UFL,BDR,BUL,RDF)

$(RU)^{-1}$ = (FR,DR,BR,UR,UB,UL,UF)$(URF)_-(BRU,LBU,FLU,RDF,RBD)_+$

$(RU)^{15}$ = (FR,UF,UL,UB,UR,BR,DR)

Problem 10. (page 8) (A,B,C,D) = (A,B)(A,C)(A,D)

Problem 11. (page 8) Referring to Problem 7, we see that I, R^2, V, H are even and R, R^3, D_1, D_2 are odd. On the cube, any basic move is a product of two 4-cycles, hence is even. Hence all possible positions are even permutations of START (considering just the positions of pieces, not their orientations).

Problem 12. (page 10) Any subgroup containing S must contain all finite products of elements of S and their inverses. But this set of finite products is closed under multiplication, has I (= a product of no terms or = PP^{-1}) and is closed under inverses, hence is already a group. Thus it is the smallest group containing S.

Problem 13. (page 10) Referring to Problem 7 and multiplying, we find RV = D_2, R^2V = H and R^3V = D_1.

Problem 14. (page 10) From the remarks on page 8. the even permutations are closed under multiplication and I is even. Further, from $PP^{-1} = I$, it follows that P is even if and only if P^{-1} is. Hence the **even** permutations are also closed under inversion and hence are a subgroup in any group of permutations. If P is an odd permutation, then PQ is odd for every even permutation Q. Since $Q \neq R$ implies $PQ \neq PR$ and since any odd permutation R can be written as PQ where Q is even, by letting $Q = P^{-1}R$, it follows that the mapping Q→PQ is one to one and onto from the even permutations onto the odd permutations. This is, there must be as many even permutations as odd permutations - if there are any odd permutations at all.

Problem 15. (page 10) Note that $(1,2)(1,3)(1,2) = (2,3)$ and $(1,i)\overline{(1,j)(1,i)} = (i,j)$, so every 2-cycle is generated by $\{(1,2), (1,3), ..., (1,n)\}$ and so every permutation is.

Problem 16. (page 10) Any even permutation is a product of an even number of 2-cycles. The product of two 2-cycles is either I, a 3-cycle or the product of two disjoint 2-cycles. By pairing off the 2-cycles in an even permutation, we see that any even permutation is a product of 3-cycles and disjoint pairs of 2-cycles. So it suffices to see that any of sets generates all 3-cycles and all pairs of disjoint 2-cycles. Now $(a,b,c)(a,b,d) = (a,d)(b,c)$ and $(a,b)(c,d) \cdot (a,b)(c,e) = (c,d,e)$ so the set of 3-cycles and the set of disjoint pairs of 2-cycles each generate the other. Now $(1,j,2) = (1,2,j)^2$ and $(1,2,j)(1,2,i)(1,j,2) = (1,i,j)$ and $(1,2,j)(1,2,i)(1,2,k)(1,i,2)(1,j,2) = (i,j,k)$, so every 3-cycle is generated by $\{(1,2,i) | i \geq 3\}$. For the last case, we must assume $n \geq 6$. We shall show that every 3-cycle $(i,n-1,n)$, $i < n-1$, is generated. By the previous case (with 1,2 replaced by n-1,n), this will suffice. We have $(1,2)(3,i) \cdot ((1,2)(3,n-1) \cdot (1,2)(3,n) \cdot (1,2)(3,i) = (i,n-1,n)$ for $4 \leq i < n-1$ and $(1,2)(3,n-1) \cdot (1,2)(3,n) = (3,n-1,n)$ and $(1,3)(2,4) \cdot (4,n-1,n) \cdot (1,3)(2,4) = (2,n-1,n)$ and $(1,3)(2,4) \cdot (3,n-1,n) \cdot (1,3)(2,4) = (1,n-1,n)$, so we are done. Exercise. **Which of these results are false for small values of n?** (Example on p 29)

5-A. The Slice Group. The subgroup where a = b = c on all six faces corresponds to moving the coordinate axes determined by the face centres with respect to the rest of the cube. This group is the same as the group of even direct symmetries of the cube and has 12 elements (though we need a bit more detail to complete the argument).

For dealing with this group, one notices that the eight corner pieces of the cube always remain in the same relation to one another. So we can consider them as fixed, with only the central slices moving. Thus we can refer to a face by its corners rather than its centre. With respect to the corner coordinates, we have three orthogonal slices which act on disjoint sets of edges. This gives us 4^3 motions of the edges. A slice acts as a 4-cycle on face centres, which is an odd direct symmetry of the face centres. However, for the edges to be lined up with the corners, each slice must be turned a multiple of 4 units and so the subgroup with a = b = c contains only even permutations of the face centres. Indeed, the same analysis as in the beginning of section 6 shows this group has $4^3 \cdot 24/2 = 768$ elements.

In the original coordinate system, try slice processes: FURF, $F^2R^2U^2$, FURFFRDF, $FURFFRDFF^2R^2U^2$. Exercises. Write each of these with respect to the corner coordinate system. **What is the maximal number of slice moves required to restore the cube to START from a pattern in this group?** (Try also R^2UF^2U. See page 31.)

5-B. The Slice-squared Group. This is a commutative group of 8 elements and every element has order 2.

5-C. Two Squares Group. This group has 12 elements. The FL, FR and RB columns are moved as units while the pairs (RU,RD) and (FU,FD) are interchanged.

5-D. The Antislice Group. Inspection of the effect of an anti-slice shows that the four edge pieces in a central layer (i.e. in a slice) remain in a slice but are rearranged. E.g. after the antislice R, the slice RL (i.e. between the R and L faces) remains unchanged, while the edges in slice FB change from UR,RD,DL,LU to FR,RB,FL,LB and the edges in the slice UD change from FR,RB,BL,LF to DR,RU,DL,LU. If we consider the slices as fixed and the face centres as moveable (somewhat like we did in the slice group), then our antislice moves carry each slice to itself. E.g. the anti-

slice R carries the edges of slice UD from: to: ,

i.e. two of the edge pieces have been interchanged by reflection. A little experiment will show that a slice can have only six possible configurations (when we fix, e.g. the FR edge). One can recognise the UD slice since it will be the only slice not containing any U's or D's. However, it is easier to consider the corners. Though the corners of a face do not all remain the same, the corners of a face display at most two colours and they will be opposite colours. Hence we can identify the R and L faces as being those which display only R and L in their corners. The RL slice will then remain between the R and L faces. Since the antislice R is the same as the antislice L, it doesn't matter which is R and which is L of the two R, L faces, nor does this affect which is the RL slice. (pp 35,54)

Try the antislice processes (in the original notation):

$(FR)^3$, $(FR)^3U^2D^2$, $FUR^{-1}F$, $(FR)^3(RF)^3$, $(FR)^3(FU)^3$. Exercise. Write each of these with respect to the above described coordinate system.

5-E. The Two Generator Group. I don't have a complete description of this group. I find that edges cannot get out of orientation, but can otherwise be moved as desired. (See pp 54-56.)

Problem 17. (page 14)

			Number of Moves
A)	i)	$FB^{-1}U(R^2U^2)^3U^{-1}BF^{-1}$	11
	ii)	$R^2D^2B^2D(F^2L^2)^3D^{-1}B^2D^2R^2$	14
	iii)	$(U^2R^2)^3B^{-1}UB(U^2R^2)^3B^{-1}U^{-1}B$	18
	iv)	$FU^{-1}B^{-1}D(R^2D^2)^3D^{-1}BUF^{-1}L^{-1}UL(F^2D^2)^3L^{-1}U^{-1}L$	25
	v)	$RBL(U^2L^2)^3L^{-1}B^{-1}R^{-1}DR^{-1}D^{-1}(L^2F^2)^3DRD^{-1}$	23
B)		$(R^2U^2)^3(U^2L^2)^3$ or $U^2D^2R^2U^2D^2L^2$	11 or 6
C)		$(F^2R^2)^3RD^{-1}RFL^{-1}F(F^2R^2)^3F^{-1}LF^{-1}R^{-1}DR^{-1}$	22

See also sections 9-A and 9-B. (See also pp 43-45.)

Problem 18. (page 16)

		Effect on edges	Moves
A)	$P_2(F,R)\cdot U\cdot P_2(R,F)\cdot U^{-1}$	(FU,RU,DR)	10
B)	$P_2(F,R)\cdot U^2\cdot P_2(R,F)\cdot U^2$	(FU,BU,DR)	10
C)	$P_3(F,R)\cdot R^{-1}\cdot P_3(R,D)\cdot R$	(FU,DF,FR)	18
D)	$P_3(F,U)\cdot$solution of C	(FU,FL,RU,DF,FR)	26

See also sections 9-B and 9-C. (See also pp 44-45.)

Problem 19. (page 16) J. H. Conway asserts his method takes less than 200 moves. (See section 9-E.) (See also pp 29,31-33,36,39-40.)

Problem 20. (page 17) R. Penrose says he has an 8 move solution but it is not clear if this is for Problem 18-A or 18-D. (See section 9-B.)

7-A. The Supergroup. The sum of all face turns in any permutation
which is I (ignoring centres) must be even since I is an even permutation
on the corners while each turn is an odd permutation of corners. Now we
have $(RU)^{105} = I$, so we have an I (ignoring centres) which turns two
adjacent centres 105 times, which is 1 unit (mod 4). Combining several
of these together, one obtains a 'centres' group of $4^6/2 = 2^{11} = 2048$
elements, which act independently of the other motions. Hence the super-

group has $2^{11} \cdot N = \dfrac{8!}{2} \dfrac{12!}{3} \cdot \dfrac{3^8}{2} \cdot \dfrac{2^{12}}{2} \cdot \dfrac{4^6}{2} = 2^{38} 3^{14} 5^3 7^2 11$

$= 885\ 80102\ 70615\ 52250\ 88000 \simeq 8.9 \times 10^{22}$ elements.
At one per microsecond, a computer would take about 2.8×10^9 years to
count through all these patterns, i.e. something like a third of the age
of the universe. (See also pp 3,18,34,38,45,46).

Basically, I haven't got much to say about the remaining problems.
The subgroup $<F^2, R^2, B^2>$ seems to have $4! \cdot 2^3/2 = 96$ elements and the
subgroup $<F^2, R^2, B^2, L^2>$ seems to have $4! \cdot 2^4/2 = 192$ elements. (See section
10-A.) The subgroup $<F, R, B, L>$ can move, e.g. FU to FR,FD,FL,UR,BR,DR,
BU,BL,BD,UL,DL but to no other orientations. Hence the F is always at F or B
and/or the U is always at R or L. So the orientation of FU can never be
changed. (See also p 32.)
J. H. Conway asserted that some element had order $2520 = 8 \cdot 9 \cdot 5 \cdot 7$, but
it doesn't seem possible for any element to have such a large order.
FRBL has order 315. (See section 10-C.)
(See section 10-D for an analysis of the commutator subgroup, 10-E for
some more pretty patterns and 10-F for some more magic polyhedra.)

In late 1978, the following UK agent was appointed: Pentangle, Over
Wallop, Stockbridge, Hampshire, SO20 8HT. They did not receive supplies
until mid-May and these were exhausted by the end of June. Further supplies
were received in late August and they will supply postal orders in the UK.
(See page 37.) Demand for the cube continues to outstrip production, even
in Hungary, despite the fact that over a million have already been made!

London, 23 Feb 1979 and 22 Sep 1979

Supplements

I have had numerous discussions and letters from many people on the
magic cube since the above material was first written. Consequently I now
have much better solutions for many of the problems and a much better
general algorithm. I am deeply indebted to: Richard Guy for giving me a
copy of the manuscript section on the cube from the forthcoming "Winning
Ways" by E. R. Berlekamp, J. H. Conway and R. K. Guy (referred to as BCG
Below); D. E. Taylor for a copy of his manuscript "The Magic Cube"; Debbie
Green and Bob Parslow for typing of a preliminary version of this revision;
Ron Mills for drawing the diagrams; Tamas Varga, Katalin Fried, Roger
Penrose, Michael Vaughn-Lee, Peter Vámos, John Gaskin and F. J. Viragh for
informative correspondence; John Conway, Roger Penrose, Chris Rowley,
Peter McMullen, Michael Vaughan-Lee and many others for informative and
enjoyable discussions.

9. SUPPLEMENT ON THE BASIC MATHEMATICAL PROBLEM. (See also pp 43-46.)

A. Improved Edge Processes.

Problem 17-A-iii. (pages 14 & 21)
$F^{-1}U^{-1}FU^{-1}F^{-1}U^2FRUR^{-1}URU^2R^{-1}$ (K. Fried - 14 moves)

<u>Problem 17-A-iv.</u> (pages 14 & 21)

$U^{-1}F^{-1}L^{-1}B^{-1}R^{-1}URBLF$ (K. Fried - 10 moves)

If we set $Q = RBLF$, then this is $U^{-1}Q^{-1}UQ$. Other combinations of Q and U produce similar 3-cycles of edges, e.g. $QUQ^{-1}U^{-1} = (UF,UR,RF)$. Another 3-cycle on these pieces is the following.

$F^2(RF)^2(R^{-1}F^{-1})^3$ $= (UF,RF,RU)$ (A. Taylor (BCG) - 11 moves)

<u>Problem 17-A-v.</u> (pages 14 & 21)

$R^{-1}U^2DB^{-1}UD^{-1}R^2U^{-1}DB^{-1}U^2D^{-1}R$ (K. Fried - 13 moves)

Some other 3-cycles of edges are the following.

$(FU)^2FU^{-1}(F^{-1}U^{-1})^2$ $= (UF,RF,UL)$ (BCG - 10 moves)

$RL^{-1}U^2LR^{-1}F^2$ $= (UF,UB,DF)$ (Varga & Fried (BCG) - 6 moves)

$RL^{-1}BLR^{-1}U^2RL^{-1}BLR^{-1}$ $= (UF,LU,RU)$ (M. Bumby (BCG) - 11 moves)

<u>Problem 17-C.</u> (pages 14 & 21)

$RL^{-1}BRL^{-1}DRL^{-1}F^2LR^{-1}DLR^{-1}BLR^{-1}U^2$ $= (UF)_+(UB)_+$ (K. Fried - 18 moves)

See section 9-B for an even better method.

Another simple edge move is the following.

$FBR(U^2B^2)^3R^{-1}B^{-1}F^{-1}$ $= (UF,LU)(UR,BU)$ (D. Singmaster - 12 moves)

B.   Monoflips and Monotwists.

Peter McMullen told me on the telephone that he had seen John Conway using processes of this type. I devised some processes of this type before I **learn**ed the details of Conway's processes from BCG. Below I describe both Conway's processes and my own, which are longer but simpler.

The basic idea is to find a process which only affects one element in a given face. Then that face can be turned and the inverse of the process is applied and the face is unturned. Thus whatever was done in the other part of the cube is undone and so only the two elements in the face are affected. BCG attributes this idea to David Seal and David Goto and uses the terms monoflip and monotwiddle.

<u>Monoflip.</u> (D. Seal) $FUD^{-1}L^2U^2D^2R = S$ flips the UF edge piece and otherwise fixes the U face, though the whole U face has been turned anti-clockwise. Thus $SUS^{-1}U^{-1}$ produces $(UF)_+(UR)_+$ in 16 moves and we can flip <u>any</u> two edges in the U face in 16 moves, thus answering Problem <u>17-C</u> more efficiently than in subsection A. (Note that the use of a monoflip or monotwist is always a commutator.) (See also pp 33,44.)

<u>Monotwist.</u> (D. Goto) $R^{-1}DRFDF^{-1} = T$ twists RFU positively and leaves the rest of the U face fixed. The negative monotwist is just T^{-1}.

$TUT^{-1}U^{-1} = (RFU)_+(RUB)_-$ in 14 moves and we can twist any two corners in the U face in 14 moves, answering Problem <u>18-C</u> in fewer moves and leaving the **edges fixed!** Using this process twice, we can twist any three or all four corners of a face in 28 (or 27) moves, answering Problem <u>18-D</u> in 4 (or 3) more moves than on page 21 (where there are 24 moves) but leaving edges fixed.

It seems difficult to improve on the monotwist, but it is perhaps worth noting that it is almost a product of commutators and $P_2(F,D) \cdot P_2(D^{-1},R^{-1})$ gives the same effect on the U face in 7 moves. (See also pp 44,46.)

Using the same idea, we note that $P_2(F,R) = FRF^{-1}R^{-1} =$ $(FLU,FUR)_+(FRD,DRB)_-(FU,FR,DR)$ only affects the FLU element of the L face. Turning L, applying P_2^{-1} and unturning L gives us

$P_2L^{-1}P_2^{-1}L = FRF^{-1}L^{-1}FR^{-1}F^{-1}L = (RFU,FLU,DLF)$ in 8 moves, which is

a solution to Problem 18-A with fewer moves and leaving the edges fixed. (Note that this appears to be 10 moves but $R^{-1}L^{-1}R = L^{-1}$ in the middle. This is apparently the solution Penrose was referring to.) Replacing P_2 by P_2^{-1} gives
$$P_2^{-1}L^{-1}P_2L = RFR^{-1}F^{-1}L^{-1}FRF^{-1}R^{-1}L = (RFU,UFL,FDL) \text{ in 10 moves.}$$
To solve Problem 18-A as stated, take $R^{-1}P_2(L,U)RP_2(U,L)$. Using the Y-commutator, we have
$$P_2(B,L^{-1})RP_2(L^{-1},B)R^{-1} = BL^{-1}B^{-1}RBLB^{-1}R^{-1} = (FLU,RUB,RFU) \text{ in 8 moves,}$$
which is a solution of Problem 18-A as stated. Indeed there seems to be much room for further investigation here as the Y-commutator affects only one piece in 3 of the cube faces and the Z-commutator affects only one piece in 2 of the cube faces. R. Penrose has found a number of these moves.

Extending the same idea, we observe that
$$P_3 = P_2^2 = (FRF^{-1}R^{-1})^2 = (FLU)_+(FUR)_+(FRD)_-(DRB)_-(FU,DR,FR) \text{ is a}$$
monotwist in the L face and a negative monotwist in the B face. This allows us to twist any 2 corners in the L face in 18 moves (actually 16 because of cancellation), giving another good answer to Problem 18-C leaving edges fixed. However, since P_3 has order 3, we can use it three times to twist any three corners in a face in 27 moves, e.g.
$$P_3L^{-1}P_3L^{-1}P_3L^2 = (LUF)_+(LFD)_+(LDB)_+, \text{ which gives a better answer to}$$
Problem 18-D. Using P_3 as a monotwist, we can twist 4 corners in a face in 32 moves, leaving edges fixed, but we can do better by noting that P_3 is also a negative monotwist in the B face. We have
$$B^2P_3B^{-1}L^{-1}P_3^{-1}LB^{-1} = (LUF)_+(LDB)_+(LBU)_-(LFD)_- \qquad (21 \text{ moves}).$$

C. Some Further Improved Corner Processes.

Problem 18-C. (pages 16,21,23,24)
$$RFR^{-1}FRF^2R^{-1}L^{-1}F^{-1}LF^{-1}L^{-1}F^2L = P_2(R,F)\cdot P_2(F^2,R)\cdot P_2(L^{-1},F^{-1})\cdot P_2(F^2,L^{-1})$$
(J. Trapp - 14 moves in two mirror image halves, leaving edges fixed)

Problem 18-D. (pages 16,21,23,24)
$$U^2R^{-1}U^2RUR^{-1}UR = (URF)_+(UFL)_+(UBR)_+(UF,UB,UR) \qquad (K. Fried - 8 moves)$$
$$FR^{-1}F^{-1}R^2U^{-1}R^{-1}U^2F^{-1}U^{-1}F = P_2(F,R^{-1})\cdot P_2(R,U^{-1})\cdot P_2(U,F^{-1})$$
$$= (FLU)+(FUR)+(FRD)+(FU)+(FR)+ \quad (M. Vaughan-Lee - 10 moves) \text{ The}$$
square of this process gives three minus twists of corners in 19 moves, thus solving Problem 18-D with edges fixed in the best way so far, and the cube gives two edge flips, thus giving another solution to Problem 17-C.
$$U^2FU^2F^{-1}U^{-1}FUF^{-1}U^{-1}FU^{-1}F^{-1}L^{-1}U^1LUL^{-1}UL \quad (K. Fried - 19 \text{ moves fixing edges})$$
Another useful corner process is the following.
$$(F^{-1}UFU^{-1}RU^{-1}R^{-1}U)^2 = [P_2(F^{-1},U)\cdot P_2(R,U^{-1})]^2 = (RUB)_+(LUF)_+(FUR)_-(FRD)_-$$
(R. Penrose - 16 moves) The two commutators involved are mirror images. Conjugating by LD gives four twists in the U face: $(RUB)_+(LBU)_+(FUR)_-(FLU)_-$ in 20 moves.

D. An Improved Algorithm.

The following is my present method though some of the above processes would shorten it considerably. I estimate that it takes at most 175 moves and I can do it in 3 to 5 minutes. It is almost entirely based on my simple processes P_1 and P_2 and so I find it easy to remember. The problem of complicated conjugation which occurred in my earlier algorithm has been eliminated.

1)   Put all bottom edges correctly in place.
2)   Put all bottom corners correctly in place.
3)   Put middle slice edges correctly in place.
4)   Flip top edges so all U faces are up.
5)   Make top orientation correct.
6)   Put top edges correctly in place.
7)   Put top corners in their right positions.
8)   Twist top corners into their correct orientations.

Generally, I carry out stages 1 and 2 on the top and then turn the cube over. It is helpful if this face has the most distinctive colour. Stage 1 is carried out by moves like F^2 (which puts FD into FU) and $F^{-1}R$ (which puts FD into UR). Stage 2 is carried out by moves like $FD^{-1}F^{-1}$ which moves BDR to URF or by combining two of these when the desired piece is in the top layer or has its U face down.

For stage 3, there are several techniques. One wants processes which affect only the top layer and one position in the middle layer. The following are some of the processes for this.

$FU(R^2U^2)^3U^{-1}F^{-1}$ moves UF to RB in 9 moves.
This is just a conjugate of $P_1(R,U)$. It is easy to determine the last two moves as being those required to restore the D face, so no memory is really required.

$FLUL^{-1}F^{-1}LU^{-1}L^{-1}$ moves UR to LF   (G. Clarke - 8 moves)
$F^{-1}U^2L^{-1}ULU^2F$ moves UF to FR   (D. E. Taylor - 7 moves)
$URU^{-1}R^{-1}U^{-1}F^{-1}UF$ = $P_2(U,R) \cdot P_2(U^{-1},F^{-1})$ moves UF to RF   (BCG - 8 moves)
    and $P_2(U^{-1},F^{-1}) \cdot P_2(U,R)$ moves UR to FR   (BCG - 8 moves)

Of course any of the solutions of Problems 17-A-iv or v and some of the other 3-cycles of edges can be used, as can the inverses of any of these. If a piece is already in the middle layer but not where it should be, it can be removed by one of these processes to the top layer. Note however that $F(R^2U^2)^3F^{-1}$ = (UF,RB)(UB,LF) moves two pieces from the middle layer, so you may be able to use it (or a simple conjugate of it) to put in two pieces at once or to put in one piece while taking out another. Alternatively, one may be able to use some of the 3-cycles of edges to move edges within the middle layer. The process $RL^{-1}U^2LR^{-1}F^2$ = (UF,UB,DF) of Varga and Fried is especially useful in this case.

For stage 4, we observe that $BP_2(L,U)B^{-1}$ operates only on the U face and flips the two pieces which were at the UF and UB positions, while $BP_2(U,L)B^{-1}$ flips the pieces which were in the UL and UB positions. So we can get all the U faces up fairly easily. (You are welcome to work out variations on this theme.) Now we can easily inspect the sides of the U edges to see if they form an even permutation. If not, apply U or U^{-1} and this is stage 5.

For stage 6, I use my solution of Problem 17-A-iii or Problem 17-A-ii.
For stage 7, I use my processes of the form $P_2(U,R)L^{-1}P_2(R,U)L$ = (UBR,LBU,LUF) or $B[P_2(L,U)]^3B^{-1}$ = (UFL,FUR)(ULB,BRU). With some observation, one can usually insure that one or two of the top corners come out in the correct orientation. For stage 8, I use my monotwist process based on $P_2^2 = P_3$.

Any process which affects just one face can be useful for stages 4,6,7,8.

Though some of the other improved processes will shorten the number of moves for the general algorithm, my method uses only the two basic processes P_1 and P_2 and most steps involve just two faces. Consequently I personally find there is little memory and little thinking time so I can carry it out rapidly. If you are willing to remember more basic processes, you can undoubtedly cut down the number of steps and perhaps speed up the time.

E.   Problem 19. BCG asserts that their method, which works from the bottom up, but in quite a different way than I have just described, takes at most 160 moves. K. Fried estimates that her method takes at most 150 moves, but I don't have details of it. R. Penrose believes that his method, which puts edges in place first, takes about 100 moves, but he hasn't analysed it in detail. It is certainly true that the average number of moves required is usually only about 50% of the maximal number. (39,40)

One can estimate a lower bound on the number of moves as follows. There are 18 first moves and then 15 second, third, ..., moves (since there is no point in turning the same face in two consecutive moves). Hence there are at most $1 + 18 + 18 \cdot 15 + 18 \cdot 15^2 + \ldots + 18 \cdot 15^{n-1}$ positions after at most n moves. This expression is equal to $1 + 18(15^n-1)/14$ and setting this equal to the number of patterns N which we saw was about 4.3×10^{19}, we obtain n = 16.60, so some patterns require at least 17 moves. We can improve on this argument slightly by considering that opposite face turns commute, e.g. FBF = F^2B. We define an <u>axial move</u> to be, e.g F^iB^j, where i,j = 0,1,2,3 but (i,j) ≠ (0,0). There are thus 15 axial moves for each axis and at most $1 + 45 + 45 \cdot 30 + 45 \cdot 30^2 + \ldots + 45 \cdot 30^{m-1}$ positions after m axial moves. Setting this equal to N yields m = 13.16, so some patterns require at least 14 axial moves. But an axial move corresponds to 24/15 = 1.6 ordinary moves on average, so we should expect some patterns to require at least 14·1.6 = 22.4, i.e. at least 23, moves. This argument could be made more determinate by considering the binomial distribution of the number of ordinary moves as a function of the number of ordinary moves. It would be most interesting if some argument could be found to improve this lower bound substantially. (See also p 34.)

10. SOME RESULTS ON FURTHER PROBLEMS.

A.   Orders of Subgroups. Let $|P, Q, \ldots|$ denote the <u>order</u> of (i.e. the number of elements in) the subgroup $\langle P, Q, \ldots \rangle$. D. E. Taylor has provided the following results. (See also pp 30,35,36,57.)

$\lvert F^2,R^2,U^2 \rvert$	=	$2592 = 2^5 3^4$
$\lvert F^2,R^2,B^2 \rvert$	=	$96 = 2^5 3$
$\lvert F^2,R^2,B^2,U^2 \rvert$	=	$1\,65888 = 2^{11} 3^4$
$\lvert F^2,R^2,B^2,L^2,U^2 \rvert$	=	$6\,63552 = 2^{13} 3^4$
$\lvert F,R \rvert$	=	$734\,83200 = 2^6 3^8 5^2 7$
$\lvert F,R,B \rvert$	=	$15999\,35016\,96000 = 2^{14} 3^{13} 5^3 7^2$
$\lvert F,R,U \rvert$	=	$17065\,97351\,42400 = 2^{18} 3^{12} 5^2 7^2$

B.   The Five Generator Group. R. Penrose's 5 generator simulation of the sixth generator can be seen by starting with, e.g. $R^2L^2U^{-1}B^2F^2U^{-1}B^2R^2B^2F^2L^2F^2U^{-1}$. This acts as though D has been used and the remaining corners are where they should be. It is fairly easy to restore the edges from this position, though doing it efficiently is tricky. One of Penrose's complete solutions is:

$$R^2L^2UF^2B^2UF^2R^2F^2B^2U^2L^2U^2L^2R^2U^2R^2U^2R^2F^2U^{-1}R^2B^2R^2L^2F^2L^2UB^2F^2U = D^{-1}$$

(31 moves). We can also see that one square is dependent on the other
five squares, e.g. $F^2B^2U^2F^2B^2R^2L^2(F^2U^2)^3R^2L^2(B^2U^2)^3 = D^2$ (21 moves).
Observing that Penrose's solution uses only U to an odd power, we ask
if $<F^2,R^2,B^2,L^2,D^2,U>$ is the whole group? (See pp 31,32.)

   C.   The Maximal Order Problem. Conway's assertion that FRBL
requires 2520 moves to return to START is based on counting FRBL as
8 moves (four turns of faces and four turns of the whole cube) rather
than as a single element of ths group. Thus he gets eight times the
order of FRBL.

   Since each **permutation** has the sum of the lengths of its edge cycles
being 12 and the sum of the lengths of its corner cycles being 8 and
since the order of a permutation is the LCM of these lengths (with
possible factors of 2 or 3 for twisted cycles), it is possible to determine just what orders can appear. A cursory examination indicates that
**the maximal order occurs when we have a 7-cycle and a twisted 2-cycle of
edges and a twisted 5-cycle and a twisted 3-cycle of corners, with the other
3 edges permuted in any twisted way.** (We cannot replace the twisted
2-cycle by a twisted 4-cycle as this gives an odd permutation on positions.)
Thus the maximum order appears to be $7 \cdot 2 \cdot 2 \cdot 5 \cdot 3 \cdot 3 = 1260$. It is easy to
see that such an element is possible, but I haven't tried to find a
sequence of moves to produce such an element. (See also pp ,49,50.)

   D.   The Commutator Subgroup. Michael Vaughan-Lee writes that the
commutator subgroup has index 2, i.e. it comprises half the whole group.
This is not too hard to see as his process,involving the product of
three commutators which is described in section 9-C, shows that any
edge flips and any **corner twists** can be achieved by using commutators.
Further, we have 3-cycles of edges either using P_2 (ignoring edges) or
using some of the solutions of Problem 17-A-iv which are commutators.
We can also get pairs of 2-cycles of corners from P_2^3 or a 3-cycle of
corners from $P_2(F,R)L^{-1}P_2(F,R)^{-1}L$. Now the conjugate of any commutator
is a commutator and the conjugate of a product is the product of the
conjugates since

$$P(QRQ^{-1}R^{-1})P^{-1} = (PQP^{-1})(PRP^{-1})(PQ^{-1}P^{-1})(PR^{-1}P^{-1})$$
$$(PQP^{-1})(PRP^{-1})(PQP^{-1})^{-1}(PRP^{-1})^{-1} \quad \text{and}$$
$$P(QR)P^{-1} = (PQP^{-1})(PRP^{-1}).$$

From all these results and our discussion in section 6, we see that we
can obtain every even permutation of edges and every even permutation of
corners by commutators. Combined with the ability to flip and twist, we
know that we can obtain half of the whole group.

   To see that we do not get the whole group, we define the <u>motion</u> of
a sequence of moves as the sum of all the exponents in the sequence, e.g
$F^2RB^{-1}L$ has motion 2+1-1+1 = 3. This motion must be considered (mod 4)
since $F^4 = I$, etc. Now the permutations of motion 0 or 2 are closed under
multiplication, so they form a subgroup. Since there are permutations of
odd motion, e.g. F, the argument of Problem 14 shows the permutations of
even motion are half the group. Since every commutator has motion 0, the
commutator subgroup is contained in the group of even motions, and since the
former is at least as big as the latter, they must be identical. (The
astute reader may **notice** that the commutators are contained in the subgroup
of permutations of motion 0, which appears to be a proper subgroup of

the permutations of even motion, since, e.g. F^2 has motion 2. However, as discussed on page 22 under the supergroup, there are permutations of motion 2 which are I (ignoring face centre turns). Hence the permutations of motion 0 and the permutations of motion 2 are identical when we ignore face centre turns. However, if one works in the supergroup, then the commutator subgroup will be precisely the patterns where both the edges and corners are evenly permuted and where the motion of each face is 0. This means the index of the commutator subgroup is 4^6 in the supergroup, i.e. it comprises 1/4096 of the group.) It follows from this analysis that F^2 and FR must be products of commutators but I haven't done this.

E.   Pretty Patterns.   Several new pretty patterns have been found.
First, consider the two 2×2×2 subcubes with outer corners at URF and DBL. These overlap in the centre of the cube and comprise 14 visible unit cubes. These can be rotated as a unit about their common diagonal while the rest of the cube remains fixed! This is a most striking pattern and I have no easy way of doing it. I apply the solution of Problem 17-A-v around URF and then its inverse about DBL to move the edges. Then the spot process of the slice group is used to move the centres and the solution of Problem 18-C is used to twist the corners. Using K. Fried's solution for Problem 17-A-v and D. Goto's monotwist, this can be done in 50 moves. Peter McMullen says he has a way to move the 'girdle' of the other 12 visible cubes by a 60° rotation but I cannot see that this is possible. (Double Cube pattern; see also pp 29,30,48.)

Another rotational pattern appears to rotate the set of corners as a unit about a main diagonal with respect to the set of edges. The centres can be turned as well. This makes patterns with six X faces or six + faces and most people will immediately think that these are in the slice group. I do this using

$$U^{-1}LP_2(F,D)U^{-1}P_2(D,F)UL^{-1}U \ = \ (FRD,UFL,RUB) \qquad (12 \text{ moves}).$$

I then apply the inverse of this about DBL and twist the centres and corners as before. This takes 52 moves. (6-X and 6+; pp 29,30,47-49)

The result of $F^2R^2L^2B^2R^2F^2B^2L^2$ is 4 vertical 'bar' faces (i.e. $a = c = b' = d'$ in Figure 13). Replacing the last L^2 by R^2 gives 4 bar faces with the bars at right angles ('crossbars' pattern). From this one can get to a number of similar simple patterns, e.g. with two X faces and four plain faces. One is also led to ask if one can form any pattern with six mutually orthogonal bar faces. One such pattern is not even constructible and another one is constructible, but involves an odd permutation on positions, so cannot be achieved in our group. There may be some such pattern which is possible. This leads to the further question (or 11 questions) of determining the pretty patterns in each of the other (11) orbits of patterns of the cube. E.g. if we disassemble the cube and reassemble with RU and RB interchanged, then a whole new set of possible patterns arises. (pp 47,48)

P. McMullen says he can rearrange the edges and centres so that all six colours show on each face. K. Fried can flip all 12 edges in 36 moves.(31,35)

F.   More Magic Polyhedra.   If your mind is not sufficiently boggled by the magic cube, you may ask if other shapes of puzzles of this nature can be made or contemplated. T. Varga says that the 2 × 2 × 2 version is in prototype. I have seen the Hungarian patent specification for the cube and it includes the 2 × 2 × 2 version, which is mechanically much trickier than the 3 × 3 × 3 version. Indeed it took several people most of an evening and models carved from corks to figure out what the diagrams meant. Varga also says the 2 x 3 x 3 (sic!) version will appear soon. (34,35) (I can't figure out what this would look like - can you?) Conceptually, the 2 × 2 × 2 version is just the corners of the 3 × 3 × 3 cube, so we

already know how to solve it. It has $8! \cdot 3^8/3 = 881\ 79840$ patterns. (31,53,60)
The edges of the $3 \times 3 \times 3$ cube have $12! \cdot 2^{12}/2 = 98\ 09952\ 76800$ patterns.

A number of variations involving fewer colours have been proposed but none seem to be of any real interest. I have recently come across a cube in which two adjacent faces accidentally received the same colour. This cube has two Oranges and no Green. Though one might think this would make the problem easier, the presence of two indistinguishable OR's and OY's and the presence of an OO mean that you cannot be sure of putting these pieces in the right places or with the right orientation until you get all the other pieces correct. Hence you either go along and then rectify things at the end or you rearrange your strategy to put these pieces in place last (which is difficult as they are not all on the same face). Further, one can move the cube into patterns which appear to be impossible by interchanging two of the indistinguishable edges or by flipping the OO piece. It takes most people some time to recognize how this has happened.

A. Taylor has shown that the group for any magic solid (apparently meaning with three elements along each edge) has the same general structure as for the $3 \times 3 \times 3$ cube, except that the total corner twist must be 0 modulo the greatest common divisor of the corner valences. (The valence of a corner is the number of edges at the corner. I cannot see how to imagine the corner pieces if different corners have different numbers of edges at them.) One can imagine the $4 \times 4 \times 4$ cube or the $3 \times 3 \times 3 \times 3$ hypercube. The first might be makeable but its group seems to be much more complicated. The second is unmakeable (?) but its group structure may be determinable. (See pp 51,52,60.)

London, 23 Sep 1979

ADDENDUM NUMBER ONE

Michael Vaughan-Lee has just sent me a detailed analysis of his general algorithm, which works by doing the edges first and takes at most 173 moves and he says one of the 3-cycles on corners would cut this down by 8 moves.(40)
25 Sep 1979

I am indebted to Richard Ahrens and Frank Barnes for further ideas and to Suzanne Lowery of The Observer and Sue Finch for the cover drawing which was originally done by Sue Finch for my article in the Observer on 17 June 1979.

Richard Ahrens has found

$$R^{-1}FRF^{-1} \cdot UF^{-1}U^{-1}F \cdot U^{-1}RUR^{-1} = P_2(R^{-1},F)P_2(U,F^{-1})P_2(U^{-1},R)$$
$$= (RFU)_+(LUF)_-(FU,UR,RF) \quad (12 \text{ moves}).$$

This rotates the RFU corner and its three adjacent edges as a unit clockwise and anti-twists the adjacent corner LUF. If we denote this move as $A(R,F,U)$, then $L \cdot A(R,F,U) \cdot L^{-1} \cdot A(D,B,L)^{-1}$ followed by slice moves BLDB (spot pattern) yields the first pretty pattern of section 10-E in 34 moves. One can also apply A to produce rotations of 4 non-adjacent corners and their adjacent edges in 48 moves. (See also p 48.)

Ahrens points out that the three Y-commutators at a corner all act only on the corner, its three adjacent edges and its three adjacent corners and he states that the 3-Y subgroup generated by these 3 Y-commutators has order 1296. We can see this because M. Vaughan-Lee's move of section 9-C is in this group and it allows us to flip and twist as desired, giving $2^3/2$ flippings and $3^4/3$ twistings. Inspection shows that we can only permute the 4 corners as pairs of 2-cycles (which illustrates the failure of Problem 16 for n = 4). There are just four of these (counting I) and they can be obtained as cubes of Y-commutators (leaving edges fixed).

Further, every commutator is an even permutation of edges (and of corners) so we can only get the three even permutations (i.e. **the three 3-cycles**) on our three edges and these can be achieved using Ahrens' move A or using a Y-commutator, though neither of these fixes the corners. Thus we have $2^2 \cdot 3^3 \cdot 4 \cdot 3 = 1296 = 2^4 3^4$ elements in this group.

Frank Barnes has also examined the Y-commutators and observes that
$$P_2(F,R^{-1})^2 \cdot P_2(U^{-1},R)^2 = (RDF)_- (RUB)_+ (UF)_+ (RF)_+ .$$

Perhaps I should note that group theorists have a standard notation for commutators: $PQP^{-1}Q^{-1} = P_2(P,Q)$ is denoted $[P,Q]$.

Ahrens has also pointed out that the 6-X pattern of section 10-E can be followed by the slice moves $F^2U^2R^2$ (the 6-X pattern in the slice group) to produce an even more confusing 6-X pattern. I have found that combining the earlier 6-X pattern with half or all of the first pattern of section 10-E produces several nice patterns, including (FL,RD,UB)(FD,RB,UL) which may be the pattern that Peter McMullen was referring to. This pattern can be achieved more directly. It consists of the 6 edges not adjacent to the RFU and LDB corners and they form a hexagon which appears to have been turned 120° with respect to the rest of the cube. Ahrens also pointed out that a 'belt' of the 6 centres and 6 connecting edges, with each edge-centre-edge triple forming a right angle, can be turned 120° with respect to the rest of the cube. This is done simply by applying the 6-spot pattern to the previous pattern. There is another way to make a 6 centre and 6 edge belt, with two edge-centre-edge triples being in a straight line, but it does not seem possible to turn this belt.

I have recently tried combining slice moves and antislice moves. Some new patterns have emerged but none are quite as interesting as previous patterns. (Central hexagon, worm, snake; see pp 31,48,49.)

Ahrens suggests "a big public competition in the manner of the great cubic solving contest of 16th century Italy. Masters of the cube could compete before the television cameras to reproduce various patterns posed by their rivals. It is too long since mathematics was a spectator sport."(38,39)

As this is now going off to be printed, let me wish you all

"Happy Cubing".

London, 2 October 1979

ADDENDUM NUMBER TWO

A number of typographical errors have been corrected. My thanks to Chris Rowley, Richard Ahrens, Lytton Jarman, Frank Barnes, Morwen Thistlethwaite, Kate Fried and Peter Neumann for pointing these out and for further information. I should also like to express my indebtedness to Trevor and Beryl Fletcher. Trevor introduced me to Tamás Varga, who gave me my first cube, at the Helsinki International Congress of Mathematicians in 1978. They also contributed a number of ideas which are contained in the main part of these notes.

While the first printing was in press, I had a visit from Dave Benson, one of Conway's Cambridge Cubists (CCC). They have made most of BCG obsolete. They have shown that the cube can always be restored in at most 94 moves. (40)

CCC are compiling an inventory of good processes for all the patterns involving a single face. There are $\frac{4! \, 4!}{2 \quad 2} \frac{2^4}{2} \frac{3^4}{3} = 62208 = 2^8 3^5$ of these which split into 288 permutations of positions and 216 changes of orientation. Using the symmetries of the face and identifying positions that differ only by a turn of the face, their inventory consists of 13 permutations of positions and 30 changes of orientation which they can do in at most 13 and 19 moves respectively, though these numbers are being constantly improved. (See also pp 32,49.)

Benson has reduced Penrose's solution of the 5 generator problem (sections 7-D and 10-C) to 13 moves. Let $A = RL^{-1}F^2B^2RL^{-1}$, then $AUA = D$. Benson also showed me how to construct more general bar patterns, where each face has three colours in three parallel bars. These can be built up by using a mono-column-flip such as $C = R^2FDR^2D^{-1}R$. Then CB^2C^{-1} gives the beginning of such a pattern. Repeat on the opposite face and then apply (FU,FD) (BU,BD) (see Problem 17-B) and then RL to complete the pattern. (48)

Richard Ahrens has shown that the corners can be rotated as a unit by $180°$ about an axis joining the midpoints of two edges. For example, take $B^{-1}P_2(R,U)^3B·U^2R^{-1}P_2(F,U)^3RU^2$. He has also shown that the second 'belt' mentioned on page 30 can be turned by $180°$. A variation of this gives the effect of interchanging the 6 edges and centres adjacent to one corner with the corresponding pieces adjacent to the opposite corner. (47-49)

Kate Fried notes that slice moves R^2UF^2U yield the 4 spot and 2 plain pattern in only 4 slice moves rather than the 8 I give on page 20. Her method of inverting all 12 edges is based on the following.

$$RL^{-1}URL^{-1}BRL^{-1}DRL^{-1}F = (FR)_+(FD)_+(BD)_+(BU)_+ \qquad \text{(12 moves)}$$

Using this three times over gives the desired result. This can be conjugated by $U^{-1}R^{-1}U$ to act only on the RL slice. (12-flip, 4-flip, see pp 28,35,47,48)

The <u>centre</u> of a group is the subgroup of elements which commute with all the elements of the group - i.e. P is in the centre if and only if $PQ = QP$ for every Q in the group. (<u>Exercise.</u> Check that the centre is a subgroup.) I asked Peter Neumann what the centre of the group of the cube was and he replied that it has two elememts: I and the pattern with all edges flipped. With a little thought, you can see that any element of the centre must act on all corners in the same way and on all edges in the same way and that there is just one such pattern other than I. (See pp 59-60.)

Several group theorists have pointed out that the interaction between the permutations of positions and the changes of orientations is an example of a <u>wreath product</u> of groups. Apparently the commutator subgroup and the centre can be determined easily from this fact. (See pp 58-60.)

Morwen Thistlethwaite has come up with a novel strategy for restoring the cube and has shown that it can always be done in at most 85 moves. (39,40)

Thistlethwaite, Barnes and others have observed that the number of patterns on the 2 × 2 × 2 cube (see page 29) can be divided by 24 to get 36 74160 since there is no fixed orientation for this cube.

London, 5 October 1979
and 22 October 1979

ADDENDUM NUMBER THREE

Many typographical errors have been corrected. Thanks to Frank O'Hara, Frank Barnes and G. S. Close for pointing these out.

My sister, Karen Rowland, writes that a silicone spray (e.g. WD-40) is a good lubricant for stiff cubes. (See also pp 1,2,37,38.)

<u>The Slice Group.</u> The 6-spot pattern (FURF on page 20) is a commutator $(FLF^{-1}L^{-1})$ in the corner coordinate system. Fried's observation shows that the 4-spot (R^2UF^2U above) is also a commutator ($R^2U^{-1}R^2U$) in the corner coordinate system. The maximal number of slice moves required to restore a given position in the slice group is at most 7 since we can get all the edges right with three slices, leaving either I, a 6-spot or a 4-spot. Frank O'Hara has shown this can be reduced to 5 and that 32 positions do require 5 slice moves.

Frank Barnes observes that the group of the cube is generated by two moves:

$$\alpha = L^2BRD^{-1}L^{-1} = (RF,RU,RB,UB,LD,LB,LU,BD,DF,FL,RD) \cdot$$
$$(FUR,UBR,LDB,LBU,DLF,BDR,DFR);$$
$$\beta = UFRUR^{-1}U^{-1}F^{-1} = (UF,UL)_+(UR)_+(UBR,UFL)_-(URF)_+.$$

Observe that α^7 is an 11-cycle of edges and α^{11} is a 7-cycle of corners, that β affects the edge and corner left fixed by α and that $\beta^2 = (UF)_+(UL)_+(UBR)_-(UFL)_-(UFR)_-$ (which is similar to Vaughan-Lee's solution of Problem 18-D on page 24). The remaining details are left as an exercise.

CCC have a program to generate all 4 move positions and compare them. I don't know if they have found anything new yet. Morwen Thistlethwaite has a program which searchs all sequences of moves from a set of faces and prints out those which move few pieces. He is up to 7 **move** sequences on 5 faces and 10 move sequences on 4 faces. Surprisingly, nothing really new has appeared, though the above process β was found. This is equivalent to my process $BP_2(L,U)B^{-1}$ (page 25, stage 4), followed or preceded by U.

I had not used this variant, but CCC had. At first, I did not believe such processes were possible as I thought $(UF,UL)_+$ must correspond to two changes of orientation. However, some thought shows that any untwisted cycle corresponds to 0 (mod 2 or 3) changes of orientations - since the first piece returns to its place after passing through all the changes - and a twisted cycle is obtained by a single flip or twist at the last position. So a twisted cycle of edges is equivalent to one flip and a twisted cycle of corners is equivalent to a twist in the direction of the subscript.

Further analysis of edge orientations resolves the problem asked in section 10-B. Consider a choice of edge orientations for the 12 edge pieces and positions. We refer to the first face and the second face of an orientation. The choices begun on page 17 can be completed as: UF, UL, UB, UR, DF, DL, DB, DR, FR, FL, BR, BL. The first faces are U or D if possible, otherwise F or B. Alternatively, the first faces are U or D and/or the second faces are R or L. We see that the moves U, D, R, L produce no changes of orientation, while F, B produce 4 changes. Hence, if a process is to produce a change of orientation (e.g. a monoflip), then it must contain an odd number of F and B moves which affect the piece. By symmetry, we can say the same for U and D and for R and L. Hence a group such as $<F^2, R^2, B^2, L, D^2, U>$ cannot be the whole group. There is a natural orientation of edges such that each 90° turn produces 4 changes of orientation. There seems to be no way to define a natural orientation on corners.

I have been inundated with algorithms and improved processes. I can only summarize them here as I have not space nor time to analyze them in detail. An article on the cube appeared in Computer Talk on 5 September and a solution, due to Colin Cairns and Dave Griffiths, appeared on 7 November. They get a face, then position the other corners, then orient them, then do the other edges. John Conway and Dave Benson (??) have prepared an article for the Journal of Recreational Mathematics (?) showing how to always restore the cube in at most 100 moves. They do bottom edges, then bottom corners and middle edges together and then use their tables to first position and then orient the top layer, in at most 13 and 19 moves respectively. (See also pp 30, 40, 49.)
Roy Nelson does bottom and middle, then orients top corners, then positions them, then orients top edges, then positions them.

Thistlethwaite's 85 move process involves first doing a 2 × 2 × 3 block, leaving say the F and R faces to do. He then correctly orients all the remaining edges (this requires using U or D), then **positions** FU, FL, FD and then puts UFL and DFL correctly in place. He then does the R edges and the R corners in at most 13 and 10 moves respectively. The technique requires some look-ahead to make sure pieces will be in acceptable places at later stages. More importantly, he has found a repertoire of 3-cycles of corners on a face which do all the possible orientation changes while doing the 3-cycles and take at most 10 moves. He can also do all reorienting 3-cycles of edges on a face in at most 10 moves. (31,36,39,44)

The most interesting new move is a new __Monoflip__ due to Frank Barnes. $W = U^{-1}FR^{-1}UF^{-1}$ flips UF but leaves the rest of the RL slice fixed. By commuting with the R slice, we can flip any two edges in the RL slice in 14 moves. (Note that the R slice moves the centres.)

$U^{-1}FR^{-1}UF^{-1} \cdot R^{-1}L \cdot DF^{-1}RD^{-1}F \cdot RL^{-1}$ = (FU)+(FD)+ solves Problem 17-C in 14 moves. Thistlethwaite modified this somehow to get:

$F^2LD^{-1}F^2B^2UR^{-1}FU^{-1}RF^2B^2L^{-1}DF$ = (FU)+(RU)+ $\qquad$ (15 moves). (p 35)

Some further shorter solutions. (See also pp 43-45.)

__Problem 17-A-i.__ (pages 14,21)

$R^2L^2DR^2L^2U^2R^2L^2DR^2L^2$ = (UF,UB)(UR,UL) $\qquad$ (R. Nelson - 11 moves)

__Problem 17-A-iii.__ (pages 14, 21, 22)

$F^2B^2DF^{-1}BR^2FB^{-1}DF^2B^2$ = (UF,UR,UB) $\qquad$ (K. Fried - 11 moves)

$R^2D^{-1}F^2RL^{-1}U^2R^{-1}LDR^2$ = $\qquad$" $\qquad$ (Thistlethwaite - 10 moves)

$R^2U^{-1}FB^{-1}R^2F^{-1}BU^{-1}R^2$ = $\qquad$" $\qquad$ ($\qquad$" $\qquad$ - 9 moves)

(The latter two are conjugates of the Varga & Fried move on page 23.)

Some other edge moves.

$B^{-1}U^{-1}BLFRUR^{-1}F^{-1}L^{-1}$ $\qquad$ = (UF,UR,BU) $\qquad$ (Thistlethwaite - 10 moves)

$F^{-1}BL^{-1}FB^{-1}U^2F^{-1}BL^{-1}FB^{-1}$ $\qquad$ = (UF,RU,UB) $\qquad$ (R. Nelson - 11 moves)

$L^{-1}B^{-1}R^{-1}URBLFU^{-1}F^{-1}$ $\qquad$ = (UF,RU,BU) $\qquad$ (Thistlethwaite - 10 moves)

$R^{-1}U^2RBLFU^2F^{-1}L^{-1}B^{-1}$ $\qquad$ = (UF,RU,UB) $\qquad$ ($\qquad$" $\qquad$ - 10 moves)

$F^{-1}R^{-1}LDF^{-1}D^{-1}RL^{-1}FU$ $\qquad$ = (FR,UB,UR) (Cairns & Griffiths - 10 moves)

$B^2R^2B^2R^2UR^2B^2R^2U^2L^{-1}RBLR^{-1}U^2B^2$ =(UF)+(UR)+(UB)+(UL)+ (Thistlethwaite -

__Problem 18-D.__ (pages 16,21,23,24) $\qquad$ 16 moves)

$(L^{-1}F^{-1}R^{-1}B^{-1})^3$ = (UFL)_(URF)_(UBR)_ · a 7-cycle and a 5-cycle of edges

$\qquad$ (W. Cutler - 12 moves)

$U^2B^2URL^{-1}B^2R^{-1}LUB^2FU^2F^{-1}U^{-1}FU^{-1}F^{-1}$ = (UFL)_(URF)_(UBR)_ (CCC - 17 moves)

$F^{-1}L^{-1}FR^{-1}F^{-1}LFRB^2L^2BR^2B^{-1}L^2BR^2B$=(UFL)-(URF)-(UBR)- (Thistlethwaite-17 moves)

(The last two both split into two interesting halves.)

$F^2D^{-1}FU^2F^{-1}DFU^2L^{-1}BLFL^{-1}B^{-1}L$=(UFL)-(URF)-(UBR)- ($\qquad$" $\qquad$ =15 moves)

$(R^2L^2DR^2L^2U^2)^2$ = (UBR,UFL)(URF,ULB) $\qquad$ (R. Nelson - 12 moves)

(If this move is repeated, acting on the D face, the result looks like the corners have been rotated as a unit by 180° about the U-D axis. This can be obtained in 14 moves using the second solution of Problem 17-B and then a slice. The 90° rotation is possible though I don't see a simple way to do it.)

I have been told of several people who have restored the cube from a random position in less than 2 minutes. The best time I have heard of is 85 seconds by Steve Rogerson. The DEC-10 at the University of Essex can do it in 2 seconds. I have been told of a sixth-form student who managed to restore a cube within half a day of first seeing one! (See pp 38,39.)

__Some problems.__ What is the shortest non-trivial identity? We have $(F^2R^2)^6$ = I = $FR^{-1}F^{-1}RUF^{-1}U^{-1}FRU^{-1}R^{-1}U$ (both of length 12) and $(F^2B^2R^2L^2)^2$ = I (length 8). Is 8 the best? (See p 36.)

One can define a figure of merit M for a move as M = LN, where L is the length of the sequence and N is the number of pieces moved. For L = 1, N = 8, so M = 8. We are only interested in the case N ≠ 0. Then the process

$P_1 = (F^2R^2)^3$ has M = 24 and the Varga-Fried move $F^2RL^{-1}U^2R^{-1}L$ has M = 18. Is this the lowest possible value?

What is the most efficient way to turn two face centres, leaving everything else fixed? (See pp 3, 18, 22, 38, 45, 46.)

Finally, I would like to acknowledge that the revision of the main part of these Notes, the addenda and the publishing were carried out while I was a Visiting Research Fellow at the Open University. I am indebted to them for the tolerance they have shown to my fanaticism.

The 2 × 3 × 3 Magic Domino is being produced. Two examples have reached the UK, though I haven't yet got one. (See below.)

London, 30 November 1979

ADDENDUM NUMBER FOUR

Again, a few typographical errors have been corrected. Thanks to John Gaskin, Frank O'Hara, Morwen Thistlethwaite and to many unnamed persons who have also found some of these errors many times over!

<u>Minimum Number of Moves.</u> O'Hara and Thistlethwaite have pointed out that my heuristic discussion of axial moves on page 26 is faulty. The sample space over which averages are taken is not clear. The average ordinary length of an axial move can be computed by considering all sequences of moves with no two consecutive moves on the same face nor three consecutive moves on the same axis. Consider any move whose predecessor is not on the same axis. Then its successor is equally likely to be any one of the other five faces, so the average length is $2 \cdot 1/5 + 1 \cdot 4/5 = 6/5 = 1.2$. Now $14 \cdot 1.2 = 16.8$, so this argument is no improvement over the original argument. There are other ways to work out this average - I have three other answers and O'Hara has two more.

We can explicitly count the number S_n of sequences of length n with no two consecutive moves on the same face nor three consecutive moves on the same axis, and we can also identify, e.g. RFB with RBF, in the process.

We have $S_0 = 1$, $S_1 = 18$, $S_2 = 12 S_1 + 27 = 243$ and $S_n = 12 S_{n-1} + 18 S_{n-2}$ for $n \geq 3$ -- since either the last move is one of the 12 moves not on the same axis as the penultimate move or the last two moves are one of the 18 pairs of axial moves not on the same axis as the antepenultimate move. (The case n = 2 requires a slight modification.) This recurrence is readily solvable by standard means and the cumulative number T_n of sequences of length ≤ n is also obtainable. Setting T_n = N yields n = 17.3, so some positions require at least 18 moves and we have made a small gain.

The values of T_n increase rapidly $(T_n \sim 1.47 (13.35)^n)$. In view of the rather low redundancy in sequences, even for the most common source, one is tempted to conjecture that every position can be achieved in at most 20 moves. This minimax value might be called the length of God's algorithm.

<u>The Magic Domino.</u> I have recieved a Magic Domino from Tamás Varga. It really does exist! It behaves like a cube with one middle slice missing — let us say the UD slice is missing. The domino group is then the same as the cube subgroup $<U,D,R^2,L^2,F^2,B^2>$ provided we identify positions differing in just the UD slice. (Group theorists say we have a quotient group.) It is easy to see that the U and D faces of pieces must always remain in the U or D directions, that is, the pieces always stay in orientation. The domino has two colours of pieces - black and white - and just the U and D faces are marked with 1 to 9 spots as on a domino. The initial position has the numbers in sequence with the whites on one face and the blacks on the other.

Now $P_1(U,R)$ yields a single 2-cycle of edges (since it interchanges a pair of non-existent edges) so we can easily obtain any permutation of the 8 edge pieces. However the commutators P_2 are not in our domino group.

But UR^2U^{-1} = (ULB,DFR)(UBR,DRB)(UB,DR)(FR,BR)
(where the last 2-cycle vanishes on the domino) affects only ULB on the
L face. Hence we can apply the same reasoning as on pages 23-24 and obtain
$$UR^2U^{-1}L^2UR^2U^{-1}L^2 = (ULB,DLF,DFR) \qquad \text{(8 moves)}$$
which is a short 3-cycle of corners both in the cube group and in the
domino group. This, together with U, shows that we can obtain any
permutation of the corners (at first affecting edges, but these can be
permuted by the previous moves). We thus have $(8!)^2 = 16257\ 02400$
positions. However, only 1/4 of these are distinguishable since we
cannot detect rotations of the domino about its UD axis. Thus we really
have $(8!)^2/4 = 4064\ 25600 = 2^{12}3^45^27^2$ positions.

Unfortunately, the fact that the even numbers are on the edges means
there is no way to arrange the numbers 1 to 9 in a Magic Square - i.e.
with all row, column and diagonal sums being equal (hence to 15). One
cannot even make all row totals the same as the first and third row totals
must be even while the second is odd. Nor can one make all the outer
row and column totals be equal. I cannot see any makeable magic pattern.
Martin Gardner suggested trying for an Antimagic Square - i.e. one with
all the totals different. I find 8 of these. (See also page 60.)

129	149	169	189	129	149	169	189
658	658	254	254	856	856	452	452
347	327	387	367	743	723	783	763

The mechanical linkage of the Magic Domino is more complicated than
for the Magic Cube. I leave it as an exercise.

Ahrens, Thistlethwaite and Benson have independently found that the
antislice group has $6144 = 2^{11}3$ elements. I am told this factors as
384 edge positions times 4 edge orientations times 4 corner arrangements.
The 8 edges in the F and B faces can all be flipped by the antislice
moves $(RFU)^2$. Further the 4 edges in the FB slice can be flipped by
$$\alpha = F^2B^2LF^2D^{-1}UR^2BL^2R^2F^{-1}L^2DU^{-1}B^2R^{-1} \qquad \text{(Thistlethwaite - 16 moves)}.$$
The product of these two flips all 12 edges in 28 or 26 moves. (One way
has a cancellation.) (8-flip, 4-flip, 12-flip. See also pp 28, 31, 47, 48.)

Thistlethwaite has shown that the 12-flip (i.e. the flip of all 12
edges) is not in the subgroup generated by slice and antislice moves.
Consider the orientation of edges with U or D first and/or R or L second,
as discussed on page 32. Consider the three edge pieces UF, UR, FR
adjacent to the UFR corner. (These can be considered as representatives
of the three slices.) Any slice or antislice move will disorient either
0 or 2 of these edges. Hence one can never get 1 or 3 edges disoriented.
This shows also that there are four edge orientation patterns,
accounting for one of the factors in the order of the antislice group.

Thistlethwaite's computer has found
$$FUF^2U^{-1}R^{-1}F^{-1}U^{-1}R^2UR = (UF,DF)_+(UR,RD)_+ \qquad \text{(10 moves)}$$
which is a twisted (perverted ?) form of P_1. From this he derives
$$R^2F^2R^2F^2RU^{-1}R^2UFRUF^2U^{-1}F = (UF)_+(UR)_+ \qquad \text{(Thistlethwaite - 14 moves)}.$$
Thus any two edges in a face can be flipped in 14 moves. He can flip any
even number of edges in the whole cube in at most 26 moves.

Thistlethwaite has found a 90° rotation of the corners as a unit
with respect to the UD axis
$$UR^2L^2DR^2L^2U^{-1}D^{-1}R^2L^2U^{-1}R^2L^2D \qquad \text{(14 moves)}$$
and his computer has found a shorter 180° rotation
$$UDLRU^2D^2LRUDF^2B^2 \qquad \text{(12 moves)}.$$
In section 5-F, I said I didn't know any simple process to interchange
one pair of edges and one pair of corners. Of course, the move β described
on page 32 does do this, but it is twisted and twists some single pieces as well.

Some neater solutions are

$UF^{-1}L^{-1}FU^2LUL^{-1}UF^{-1}LF$ = (UF,UR)(URF,LBU) (Thistlethwaite - 12 moves)
and the conjugate of this by B which yields (UF,UR)(URF,UBR) in 14 moves.

   Thistlethwaite has reduced his strategy to a maximum of 80 moves.
He has a new strategy for which the maximum has not been determined.
This has three stages: get edges into their slice with correct orientation;
move corners so as to get into the square group $<F^2,B^2,U^2,D^2,R^2,L^2>$;
use slice and antislice moves to get to START. He believes the maxima
for the stages are 10, 16 and 15 respectively. (See also pp 31-33,39.)

   The square group is contained in the group <A,S> generated by slices
and antislices and comprises 1/24 of it.

$|A,S|$ $=$ 159 25248 $= 2^{16}3^5$

$|L,R,F^2,B^2,U^2,D^2|$ $=$ 1 95084 28800 $= 2^{16}3^5 5^2 7^2$

   Thistlethwaite's computations show that the shortest identities are of
length 8 and are all instances of the commutativity of the slice-squared
group. There are no identities of length 9 and lots of length 10. (p 33)

   Two general techniques of process construction might be noted here.
One can find processes which fix corners by using slice and ordinary moves,

say S and M, in the form of a commutator $SMS^{-1}M^{-1}$, provided we use
coordinates such that a slice move is considered as a rotation of the
centre layer with respect to the coordinate system. In these coordinates,

the Varga-Fried move (page 23) is of the form $(RL^{-1})F^2(R^{-1}L)F^2$ since the
R slice move carries the U centre to the F face. In simple cases, these
coordinates are defined by the corners, but in more complicated cases it
is easy to lose track of the coordinates so I have generally avoided using
this system. Slice moves have no effect on the corners in this system,
so the effects of M and M^{-1} cancel. The second solution of Problem 17-B
(page 21) is of the form $(U^2D^2)L^2(U^2D^2)L^2$ in these coordinates and K. Fried's
4-flip is $((RL^{-1})U)^4$. Numerous other processes may be expressed more simply
or discovered in this system - particularly those with the corners fixed.

   If we have a process such as RULB, then ULBR = R^{-1}(RULB)R is a conjugate
of it and so it has the same cycle structure as RULB. This is an easy way
to rearrange a process and to look for useful variations, especially to
obtain cancellations in combining with other processes. Note that the
shift can be repeated, i.e. we also look at LBRU and BRUL. (See also p 60.)

   There are several ways to form a reflection of a process. A few of
these have been used but I don't know of any systematic study of such
processes.

   Pentangle (page 22) is now distributing a summary algorithm due to
G. S. Howlett which uses the Y-commutator (page 17), the move β described
on page 32, their powers and simple conjugation. Abbreviated versions of
my second algorithm of section 9-D have been prepared by John Gaskin and
Richard Maddison. (And now by myself as A Step by Step Solution of Rubik's
'Magic Cube', pp S1-S4 of these Notes.)

London, 16 January 1980
Slightly revised on 3 August 1980

ADDENDUM NUMBER FIVE

   A number of new features have been added to this version, as explained in the new preface. Once again, a fair number of errors have
been corrected - many thanks to Michael Holroyd, Roland J. Lees, Frank
O'Hara and especially to David C. Broughton who checked all the processes
with a computer program. I have added a lot of cross references in the
preceding text and in the present Addendum.

Many thanks also to the following for useful communication:
R. Ager, Peter Andrews, J. D. Beasley, Dave Benson, Tom Brown, Joe Buhler,
J. B. Butler, Uldis Celmins, Paul Coates, Marston D. E. Conder,
Krystyna Dałek, Charlotte Franklin, Alexander H. Frey, Jr., Kate Fried,
John Gaskin, Ron Graham, Richard K. Guy, Nicholas J. Hammond,
Guy Haworth, Michael Holroyd, Gerzson Kéri, Gil Lamb, Charles Leedham-Green,
Roland J. Lees, Steven Mai, Bill McKeeman, Kersten Meier, Susan Mills,
Roberto Minio, Jane Nankivell, Peter Neumann, Frank O'Hara,
Dame Kathleen Ollerenshaw, Zoltán Perjés, Oliver Pretzel, Steve Rogerson,
Chris Rowley, Itsuo Sakane, Sam L. Savage, Zbigniew Semadeni, Peter Strain,
Bela J. Szalai, Don E. Taylor, Paul B. Taylor, Morwen B. Thistlethwaite,
John Trapp, Trevor Truran, Peter Vámos, Tamás Varga, Michael Vaughan-Lee,
Richard Walker. I should especially like to thank Krystyna and Jurek Dałek
for arranging a visit to the Universities of Warsaw and Gdańsk during which
I started work on this edition. Ron Mills has again provided diagrams.

5.1. LUBRICATION.

Steven Mai says that French chalk is a good lubricant for the cube.
Others and I have found that WD-40 can make the cube and one's hands a
bit tacky. I am told that silicone grease or other silicone sprays may
be better.

5.2. DISTRIBUTION OF THE CUBE.

As of early 1980, a large US firm, Ideal Toy Co., has acquired the
distribution rights for most of the western world. Ideal has subsidized
a number of improvements in the manufacture and new packaging. The new
version is now being sold in the US at prices in the $7.50 to $15.00 range,
based on a wholesale price of about $5.00. However the UK wholesale price
is about £4.00 (including 15% VAT) and the retail price is £6.00 to £10.00.
Ideal has renamed the cube as "Rubik's Cube" on the grounds that
'magic' tends to be associated with magic, as in magicians. Consequently,
I have retitled these Notes. Ideal is promoting the cube in the US with
TV ads, tee-shirts, Rubik and Zsa-Zsa Gabor (who must be among the least
cubical of Hungarian exports!). Meanwhile production of the Magic Domino
seems to have been suspended.
Logical Games, Inc., 4509 Martinwood Drive, Haymarket, Virginia, 22069,
USA, tel: 703-754 4548, has set up a US production of the cube, using white
plastic. They are selling them at $9.00 postpaid in the US. Postage to
Europe is $1.00 by sea and $2.00 by air. Their wholesale price is about
$5.00.
Pentangle (address on page 22, tel: 0264 78-481) now supplies cubes,
as an agent of Ideal, for £6.00 in the UK, £9.00 in Europe and £10.00
elsewhere, all postpaid. Optikos, 12 Heath Grove, Buxton, Derbyshire,
SK17 9EH, UK, tel: 0298-2779, will also supply at similar prices.
I am told that the cube is being made in Hong Kong and will sell for
1980 yen (approx. $9.00) in Japan.

5.3. INDEPENDENT INVENTION.

Itsuo Sakane, an arts and sciences correspondent for Asahi Shimbun, in-
forms me that Terutoshi Ishige, a self-taught engineer and owner of a small
ironworks about 100 km east of Tokyo, has also invented the Magic Cube.
This was about five years ago, first in the 2 × 2 × 2 form and then in
the 3 × 3 × 3 form with two somewhat different mechanisms. Sakane has
kindly sent me copies of the three Japanese patents, which show first
(filing?) dates of 1976 and 1977, second (issuing?) dates of 1978 and
third (publication?) dates of 1980. Rubik's Hungarian patent shows three
dates of 1975, 1976 and 1977, so he seems to have priority. Ishige's
patent drawings show mechanisms conceptually equivalent to Rubik's, but

quite different in details. To me, Ishige's designs appear less simple and less robust than Rubik's but I feel that further development would have led to something like Rubik's.

James Dalgety, of Pentangle, and I have received a number of other independent designs for the cube. Some of these are just reinventions of Rubik's design and none of the others seems sufficiently practicable.

5.4. GENERAL ANECDOTES.

A colleague of mine, Paul Taylor, while walking in the South Downs at Easter, found a cube in a pub with a bottle of Scotch offered to anyone who could restore it. He managed to do it after a while, but the landlord accused him of cheating (magic ?) and refused to pay up!

A Los Angeles department store offered $50 to anyone who could put one face right in three minutes. They lost $600 to a group (sic!) of Don Goldberg's students at Occidental College.

Several reports tell me that John Conway now restores the cube behind his back, with only four or five 'looks'.

Barratt Developments Ltd., a major UK home builder based in Newcastle upon Tyne, printed two of their company logos on opposite faces of cubes and sent them as invitation to the press reception and opening of their showhouse at the Ideal Home Exhibition in London on 3 March 1980. The printing makes it essential that the two centres be correctly oriented (see p 18, 22). That is, we must work in the supergroup. Processes for centre moves will be given in section 5.8-F.

Dame Kathleen Ollerenshaw, lately Lord Mayor of Manchester and a well known recreational mathematician, writes that she has developed 'cubist's thumb' - a form of tendonitis requiring minor but delicate surgery for its relief. It has afflicted her left thumb even though she is right-handed. The problem seems to occur most commonly among teenage disco freaks as a result of too much **finger**-snapping, but it also occurs in over-enthusiastic users of (non-electric) hedge clippers, and it used to be common in car starting handle crankers. It also afflicts horse's fetlocks.

Kate Fried says she has been able to invert up to 7 random moves from START. This is the basis of a good exercise or a two person competition. One person makes a given number, say 6, moves from START and gives the cube to the other who must get back to START in the given number of moves.

5.5. COMPETITIONS AND SHORTEST TIMES.

In early 1980, Kate Fried organized cube competitions in Budapest, at the Youth Mathematical Circle (Ifjúsági Matematikai Kör) and at a Magic Cube Fans Club. Each competitor provides his own cube in its cardboard box with his name on it. (The original Hungarian packaging is an ingenious blue cardboard cubical box. Some similar boxes were used in Germany and the UK.) The judge(s) then randomize the cubes identically and put them back in their boxes. In the first competition, on 4 January, a student named Viktor Tóth won in 55 seconds. After this competition, the winner and the runner-up had a play-off or second dimension competition in which each randomized the other's cube. Tóth won again.

For the second competition, in March, over a thousand people came, overflowing several rooms and into the street. A request for persons who could do the cube in less than 90 seconds yielded nine people and a competition was held among them. The winner did his in 40 seconds.

A number of UK cubists have become 'speed merchants'. This involves hours of careful filing, greasing and oiling to obtain a cube of optimum smoothness. Peter Strain told me he had achieved 46 seconds just a few days after I had heard of the 40 second record. He also said that he had counted a total of 6704 moves in a hundred restorations. Nicholas J. Hammond says he has achieved 36 seconds, but as he admits, this was a bit

of a fluke as the cube was nearly correct when the first layer was done.
Also, he allows himself some viewing time before starting. I think this
makes it difficult to compare with competition situations and I suggest
that viewing time ought not to be allowed. Hammond insists that single
times are not a true measure of speed and suggests taking the total time
for ten consecutive restorations. His best average for ten is 54.7 seconds!
He reports that a smooth cube can be turned 100 to 120 times in a minute.
(He does FRBLFRBL...)

J. B. Butler says his computer can restore a cube in $\frac{1}{2}$ second of
CP time. (See also p 33.) I don't really know if it is meaningful to
compare computer times of this sort of length.

5.6.   PROBLEM 19 - MAXIMUM NUMBER OF MOVES TO RESTORE THE CUBE.

(See pp 16, 26, 29-32, 36 for earlier work.)

A.   THISTLETHWAITE'S ALGORITHM.

Morwen B. Thistlethwaite, henceforth denoted by MBT, shares my office
at the Polytechnic. He has been the most diligent user of computing to
find efficient processes and algorithms. His current algorithm requires
at most 52 moves, but he hopes to get it down to 50 with a bit more computing
and he believes it may be reducible to 45 with a lot of searching. The
method involves working through the following sequence of groups:

$$G_0 = \langle L, R, F, B, U, D\rangle;$$
$$G_1 = \langle L, R, F, B, U^2, D^2\rangle;$$
$$G_2 = \langle L, R, F^2, B^2, U^2, D^2\rangle;$$
$$G_3 = \langle L^2, R^2, F^2, B^2, U^2, D^2\rangle$$
$$G_4 = \{I\}.$$

Once in G_i, one only uses moves in G_i to get into G_{i+1}. The ratio $|G_i|/|G_{i+1}|$
is called the index of G_{i+1} in G_i. The indices involved are:
2048;   10 82565;   29400;   6 63552 $= 2^{11}$; $3^7 \cdot 3^2 5 \cdot 11$; $2^3 3 \cdot 5^2 7^2$; $2^{13} 3^3$.
The factors of 2^{11} and 3^7 in the first two indices correspond to the fact
that these stages rectify the orientations of edges and corners. The present
state of knowledge is summarized as follows, where stage i is the step from
G_{i-1} to G_i.

Maximum number of moves	STAGE				
	1	2	3	4	TOTAL
Proven	7	13	15	17	52
Anticipated	7	12	14?	17	50?
Best possible	7	10?	13?	15?	45?

The results for stages 3 and 4 require extensive tables comprising about
500 and 172 entries respectively as well as some preliminary reductions.
However, MBT has told me today that he has reorganized stage 3 and that
it no longer requires such extensive tables.

An earlier MBT algorithm proceeded to first orient the edges and get
them into their slices (18 moves) by a process similar to the above algorithm.
The edges were then put in place (9 moves) and the corners were done
(36 moves) making a total of 63 moves at most.

B.   OTHER ALGORITHMS.

I have received a large number of algorithms in various stages of prep-
aration and publication. I summarize them below. Bibliographic details are

in the Bibliography (pages B1,B2), or else the item is a private
communication.

James Angevine's solution is distributed by Logical Games (page 37).
It proceeds in the same sequence as Cairns and Griffiths (page 32).

J. D. Beasley sends a 'downmarket' solution, i.e. with no mathematical notation. He does corners, then edges, then centres!

The draft article described as CCC on page 32 is by Dave Benson,
John Conway and David Seal. I shall denote it as BCS. Most of the
previous attributions to CCC are to early unsigned versions of BCS.
Richard Guy has sent a recent revision of BCG. This gives a number of
new ideas and says the BCS method is down to 85 moves.

Hanke Bremer did a solution without having a cube. He gets corners
in place, then oriented, then edges in place, then oriented. Joe Buhler
has produced a solution which is being distributed in the US but I haven't
seen it. Michel Dauphin, a sixth year student at a lycée in Luxembourg,
gives a method similar to Bremer's.

3-D Jackson (apparently 3-D is a pseudonym for Bradley W.) has compiled
"The Cube Dictionary" giving about 177 one layer processes and 16 edge(s)-
into-middle-slice processes. He asserts these allow the following two
algorithms. I: D edges (10); D corners (22); middle edges (28); U corners
in place (8), then oriented (14); U edges (10) making a total of 92 moves.
II: a 2 × 2 × 3 block (30); a remaining face (20); the other face (20)
making a total of at most 70 moves. (Compare with MBT, page 32.)

Gerzson Kéri says he can do two layers in at most 57 moves and that
he believes any one layer process can be done in at most 18 moves (see
section 5.10-A). He says he has over 100 one layer processes of at most
15 moves. He has published a lengthy two-part article in Hungary and
apparently the manufacturers objected to his making the solution too clear!
One of his methods is entirely tabular - proceeding one piece at a time,
you see where it is and then look up a process to get it correctly in place.
This does D corners, U corners, FB slice edges, RL slice edges, then UD slice
edges. The table contains 24 + 21 + ... + 3 + 24 + 22 + ... + 2 = 264
entries of which 5 are impossible and 20 are trivial. This takes at most
182 moves.

Bill McKeeman sends an algorithm due to Adam Kertesz which does two
layers, then U corners in place, then U edges in place, then orients U edges
and then U corners. Kersten Meier sends a layer by layer algorithm but
doesn't detail the working for the last layer. Dame Kathleen Ollerenshaw's
method does the D face, then U corners, then middle slice edges, then U edges.
She reports an average of 80 moves.

Zoltán Perjés has written out an algorithm, based on Penrose's (page 26),
which does D edges, middle slice edges, U edges, then corners in place and
then oriented. He also outlines Rubik's original fast method: D corners,
U corners in place, then oriented, three D edges, three U edges, the other
D and U edges, then the middle slice edges.

Don Taylor does D edges, D corners, middle edges, U corners in place,
U edges in place, U corners oriented, then U edges oriented.

Michael Vaughan-Lee has made a careful study of an edge first process:
D edges (13), three middle edges (14), U edges and last middle edge (17)
(hence 44 for the edges), corners in place (30) and then in orientation (38),
but the last two steps can be reduced to 26 + 24, 32 + 22 or 28 + 22,
giving a maximum of 98 moves.

5.7. NEW NOTATION AND DIAGRAMS.

Several people have suggested that R^{-1} be written R'. I had hesitated
at doing this since I sometimes use P' to denote another permutation than P.
However there seems to be little danger of confusion and it is much easier
and neater to use ', so we shall adopt it henceforth. I shall use $^{-1}$ in a
few situations.

I shall write URF$_+$ for (URF)$_+$, etc.

After several thoughts about ways of writing slice and antislice moves within ordinary processes, I have decided to write:

$$R_s = L'_s \quad \text{for } RL' = L'R;$$
$$R_a = L_a \quad \text{for } RL = LR;$$
$$R'_a = L'_a \quad \text{for } R'L' = L'R'.$$
$$R^2_s = R^2_a = L^2_s = L^2_a \quad \text{for } R^2L^2 = L^2R^2.$$

Clearly s stands for <u>s</u>lice and a for <u>a</u>ntislice. <u>WARNING.</u> These are normally taken with respect to <u>our</u> ordinary centre coordinates, <u>not</u> the corner coordinates (pp 20-21) <u>nor</u> the fixed spatial coordinates described on page 36.

I shall also use $[\overline{F,R}]$ for $P_2(\mathbf{F},\mathbf{R}) = FRF'R'$ on occasion (page 30).

Various notations have been used for computer output, e.g. R', R**3**, r. Several people have asked for a notation to represent physical turning of the cube. I find this is only needed when one wants to apply the same process at different places on the cube within a large process, e.g. with P_1, P_2 or Ahrens' A (p 29), etc. I have not felt sufficient need to set up such a notation. Zbigniew Semadeni points out that FR can mean a piece, a position or a process and that this can get confusing. I have generally not found this to be a difficulty but one must be careful in writing out an elementary algorithm. You **are** welcome to use different styles of letters for the different uses, but I will just try to be careful of my language when necessary.

Dame Kathleen Ollerenshaw uses the diagram of Figure 5.1 with a separate 3 x 3 square for the D face. In most processes, the D face is unchanged, so it is normally omitted. <u>Exercise.</u> If only Figure 5.1 is filled in, how many possible D faces are there? (I get 12, 2 or 1.) The diagram of Figure 5.2 came from an unknown source. This shows the whole cube except for one centre. Though both of these are elegant, I find them too complex to work with.

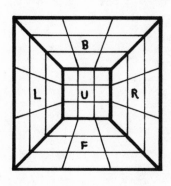

FIGURE 5.1

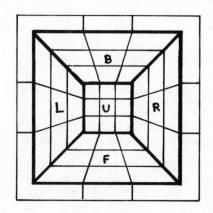

FIGURE 5.2

A. U NOTATION.

I have recently been using a simplified diagram and notation for dealing with one (face) layer processes, based on BCS, MBT, 3-D Jackson and Ollerenshaw's work. Since I always put one layer processes in the U face, I will call them U processes to indicate their superiority over non-U processes. The following will be called U diagrams and U notation.

Consider a typical U process:
P = BLUL'U'B' = B[L,U]B' = (ULB,UBR)$_+$(UFL,URF)$_-$(UF,BU,UL).

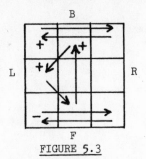

FIGURE 5.3

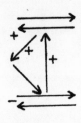

FIGURE 5.4

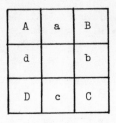

FIGURE 5.5

I diagram this as in Figure 5.3. The figure shows only the U face and
I usually omit the F, R, B, L around the edges which are then assumed.
The arrows show the change of position. A sign by an arrow head indicates
the change of orientation of the piece moved by the arrow to the position
where the arrow head is. If we begin at START, these changes of orientation
are easily seen by just looking for the U faces of each piece. Thus the
UF piece goes to the UB position and the + shows that its orientation is
changed, i.e. it goes to BU in our usual notation. The U diagram can be
drawn without the 3 × 3 grid as in Figure 5.4. One can put the sign at
the middle of the arrow rather than by its head. For complicated processes,
one may draw the corner and edge diagrams separately. Pieces which remain
in place but are flipped or twisted have just a sign shown at the position.

It also helps to have an abbreviated notation for pieces and effects
of U processes as is done in 3-D Jackson's Dictionary. He numbers both
the corners and edges from 1 to 4 and then he must specify whether (1,2,3)
is an edge 3-cycle or a corner 3-cycle. I think two kinds of symbols is
better and I will use the the lettering shown in Figure 5.5. (The corners
are the same as my canonical example of the symmetries of the square (p 10).
Signs are used to represent flipped or twisted orientations. Thus we can
write

$$P = BLUL'U'B' = (ULB,UBR)_+(UFL,URF)_-(UF,BU,UL)$$
$$\text{as} \quad (A,B)_+ \quad (D,C)_- \quad (c,a+,d)$$
$$\text{or} \quad (A,B)_+ \quad (C,D-)_- \quad (a,d+,c+)$$

corresponding to $(ULB,UBR)_+(URF,LUF)_-(UB,LU,FU)$.
Note that in this notation, as in our original notation, a cycle records
the sequence of images of its first element. This makes it easy to read
off the powers of the permutation, e.g. $P^2 = A_+B_+C_-D_-(a,c+,d+)$. However,
it is sometimes more informative to see the change each piece undergoes
as is shown in the above U diagrams. For example, we have a → d+ which
shows that the piece at a will be flipped in going to d, but d+ → c+
shows that the piece at d is <u>not</u> flipped in going to c. The signs shown
in the U diagrams can thus be considered as the <u>differences</u> between the
signs in the U notation. Note that the sum of the arrow signs in a cycle
gives the twist of the cycle.

The two usages of the signs are perhaps a bit confusing at first, but
both are useful in different contexts so it is
best to understand both. The two usages of
signs behave slightly differently under
inversion. For example, P' is diagrammed
in Figure 5.6 and we have
$$P' = (B,A)_-(D-,C)_+(c+,d+,a)$$
$$= (A,B-)_-(C,D)_+(a,c+,d+).$$
In the diagram, arrows are reversed and
signs are inverted, where the inverse of a
twist is a twist of the opposite sign but
the inverse of a flip is still a flip. The

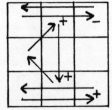

FIGURE 5.6

signs are on the same arrows as before, so one may want to place them at the middle of the arrows so that they are not moved by the inversion. In the cycle notation the cycles are all reversed and the twists are inverted, but the orientation signs on positions are not changed. Once the inverse is computed in this way, one can easily shift the cycles to get any desired element first. E.g. $(A,B+,C-)_+^{-1} = (C-, B+,A)_- = (B+,A,C+)_-$ $= (A,C+,B)_-$. Exercise. Translate the processes of Problems 17-A, 18-A, 18-B, 18-D into U notation and draw U diagrams for each. Also do all their inverses.

5.8.   A SMALL CATALOGUE OF PROCESSES.

I have received a great many letters giving improved processes. Sometimes the same or equivalent processes have come from several people so I have no way of attributing them to any one person. I will attribute these to 'well known', denoted by WK. I shall abbreviate the most common sources as follows.

BCS	Benson, Conway & Seal as quoted in BCG
BCG	Berlekamp, Conway & Guy
3DJ	3-D Jackson
GK	Gerzson Kéri
KO	Dame Kathleen Ollerenshaw
DBS	David Singmaster
MBT	Morwen Thistlethwaite
MBTC	Morwen Thistlethwaite's computer

My apologies to anyone who feels that he has not been credited with his invention, but most of these processes have probably been discovered by many people. An attribution such as (WK - 10; 17-A-i, pp 14,21,33) means that the process is well known, it takes 10 moves, it is a solution to Problem 17-A-i (or closely related) and that it (or close relatives) are on pages 14, 21 and 33 of these Notes.

When there are several equivalent processes with the same result, I usually give only one of them, chosen for convenience of representation or execution. There are now so many processes that it is necessary to impose some order and some selectivity on them. I have begun on this below. I will be concentrating on U processes, on edge processes and on corner processes. I give complete repertoires of U 3-cycles on edges and on corners and of U flips and U twists. I list the cycle structure first for convenience of use and I have tried to include all non-obsolete processes from earlier in these Notes. After the catalogue, I will discuss some interesting new methods.

Note, of course, that use of U, U^2, U' and/or physical rotation of the cube may simplify a given U position into one of those catalogued.

A.   U EDGE PROCESSES.

(i) Two 2-cycles.

There are 16 possible cases here, which can be reduced to 10 by symmetry and inversion, but not all have nice known solutions.

$(a,b)(c,d)$	$= F_s^2 D' L^2 F_s^2 R^2 F_s^2 DF_s^2$	(R. Walker & 3DJ - 12; 17-A-ii, pp 14,21)
$(a,b+)(c,d)$	$= RBUB'U'R^2F'U'FUR$	(D. E. Taylor & 3DJ - 11)
$(a,b+)(c,d+)$	$= F_aR(U^2B^2)3R'F_a'$	(DBS - 12; p 23)
$(a,c)(b,d)$	$= R_aU^2R_a' \cdot F_a'U^2F_a$	(WK - 10; 17-A-i, pp 14,21,23)
$(a,c)+(b,d)+$	$= F'U'L'U^2LFURU^2R'$	(3DJ - 10)

(ii)   One 3-cycle.

There are 4 possible cases here. I write them in the form (a,c,b) since that is the form used in 17-A-iii and on p 33. The triple of signs shows whether the corresponding piece is flipped as it is moved, i.e. they are the arrow signs of the U diagram.

(a,c,b)	= 000 =	$R^2U'F_sR^2F'U'R^2$	(MBT & WK - 9; <u>17-A-iii</u>, pp 14,21,22,33
$(a,c,b+)$	= 0++ =	$RD^2L^2BL^2D^2R^2R^2FR$	(R. Walker - 9; p 33)
$(a,c+,b)$	= ++0 =	$L'B'R'URBLFU'F'$	(MBT - 10; p 33)
		$L'[R,U]L[R',F]$	(O. Pretzel - 10; p 33)
$(a,c+,b+)$	= +0+ =	$B'U'BLFRUR'F'L'$	(MBT - 10, p 33)

(iii) Flips.

There are 3 possible cases here. The quadruple of signs shows
the effect on a, b, c, d.

a_+b_+	= ++00 =	$B'U^2B^2UB'U'B'U^2FRBR'F'$	(MBTC - 13; pp 23,24,35)
a_+c_+	= +0+0 =	$U'FR'UF'R_sUB'RU'BR'_s$	(F. Barnes - 14; <u>17-C</u>, pp 14,)
	=	$LF'UL'F_s \cdot U \cdot R'FU'RF'_s \cdot U'$	(BCS - 14; 21,23,24,33)
$a_+b_+c_+d_+$	= ++++ =	$R^2B^2R^2U^2R_s \cdot B \cdot R'_sU^2R^2B^2R^2 \cdot U$	(MBT - 14; pp 31,35)

B.   U CORNER PROCESSES.

(i)   Two 2-cycles.

There are 54 possible cases here which can be reduced to 18 by
symmetry and inversion. Surprisingly few of these have come up simply.

$(A,B+)(C,D+)$	=	$B[L,U]^3B'$	(DBS - 14; p 25)
	=	$(FDF^2D^2F^2D'F'U^2)^2$	(BCS - 16; p 25)
$(A,B)_+(C,D-)_-$	=	$R'F'RL^2B'R'BL^2F_sRB$	(MBT - 12)
$(A,C)(B,D)$	=	$R_aU^2R' \cdot F'U^2F_a \cdot U^2$	(N. J. Hammond & BCS - 11; p 33)
	=	$R^2F^2_sL^2D \cdot R^2F^2_sL^2U$	(R.Walker - 10; p 33
$(A,C)_+(B,D+)_-$	=	$B'U'BU'BUB^2UB^2U^2U^2B'U^2$	(MBT - 12)

(ii)   One 3-cycle.

There are 9 possible cases here. I write them in the form (A,D,B)
since that is the form of my favorite 3-cycle (p 25), though it is inverse
to the form of Problem 18-A (pp 16,21,24). The triple of signs shows how
the corresponding piece is twisted as it is moved.

(A,D,B)	= 000 =	$B^2L^2BRB'L^2BR'B = B^2[L^2,[B,R]]B^2$	(MBT - 9)
	=	$LF'LB^2L'FLB^2L^2 = L^2B^2[B^2,[L',F']]B^2L^2$	(GK - 9)
$(A,D,B+)$	= 0+- =	$LB'D^2BU^2B'D^2BU^2L' = LB'[D^2,[B,U^2]]BL'$	(MBT - 10)
$(A,D,B-)$	= 0-+ =	$R'FRB'R'F'RB = [[R',F],B']$	(3DJ - 8)
	=	$BLFL'B'L'F'L' = [B,[L,F]]$	(KO - 8)
$(A,D+,B)$	= +-0 =	$LFR'F'L'FRF' = [L,[F,R']]$	(3DJ & GK - 8; p 24)
	=	$B'R'BL'B'RBL = [[B',R'],L']$	(MBT & KO - 8; p 24)
$(A,D+,B+)$	= +0- =	$R^2F^2R'B^2RF^2R'B^2R' = R[R,F^2],B^2]R'$	(MBT & KO - 8)
$(A,D+,B-)$	= +++ =	$URU'L'UR'U'L = [[U,R],L']$	(DBS & WK - 8; pp 23,25)
	=	$BU'F'UB'U'FU = [B,[U',F']]$	(DBS - 8; pp 23,25)
$(A,D-,B)$	= -+0 =	$B^2D'BU^2B'DBU^2B = B^2[D',[B,U^2]]B^2$	(GK - 9)
$(A,D-,B+)$	= --- =	$FL'B^2L'F^2LB^2L'F^2L^2F' = FL'[B^2,[L',F^2]]LF'$	(DBS - 11)
	=	$BD'B^2D'F^2DB^2D'F^2D^2B' = BD^2[[D,B^2],F]D^2B$	(KO - 11)
$(A,D-,B-)$	= -0+ =	$F'L^2F'R^2FL^2F'R^2F2 = F'[L^2,[F',R^2]]F$	(MBT & KO - 9)

(iii)   Twists.

There are five possible cases here. The quadruple of signs shows the
effects on A, B, C, D.

A_+B_-	= +-00 =	$L(U^2LB'D^2BL')^2L'$	(E. Rubik & WK - 13; <u>18-C</u>, pp 16,21,23,24)
A_+C_-	= +0-0 =	$(U^2BR'D^2RB')^2$	(E. Rubik & WK - 12; pp 23,24)
$A_+B_+C_+$	= +++0 =	$U^2LF'L^2FLF'L^2FU^2BLB'L'$	(MBT -14; 18-D, pp 16,21,23,24,32,33)
$A_+B_+C_-D_-$	= ++-- =	$L'FD^2LF^2D'F \cdot U^2 \cdot F'DF^2L'D^2F'L \cdot U^2$	(MBT - 16; pp 23,24)
$A_+B_-C_+D_-$	= +-+- =	$L(FU'RUR'UF'U')^2L'$	(BCS - 18; pp 23,24)

C.   SOME OTHER U PROCESSES.

Below is a selection of U processes which are particularly short or simple.

$(A,B)(a,b)$	$=$	$F'UBU'FU^2B'UBU^2B'$	(3DJ & MBT - 11; p 36)
$(A,B-)(a,d)$	$=$	$R'URU^2R'URU'LU^2$	(J. Trapp - 11)
		a	
$(A,C+)(a,d)$	$=$	$FRUR'F^2LFL^2ULU'$	(GK - 11; p 36)
$(A,C)_-B_+a_+(b,c)_+$	$=$	$URBUB'U'R' = UR[B,U]R'$	(F. Barnes & BCS - 7; p 32)
$(A,B)_+(C,D-)_-(a,b,c+)$	$=$	$F[U,R]F'$	(DBS - 6)
$(A,B)_+(C,D-)_-(a,d+,c+)$	$=$	$B[L,U]B'$	(DBS - 6; p 25)
$(A,C)_+(B,D+)_-(a,c,b)$	$=$	$[F,U][U^2,F]$	(J. Trapp - 7)
$(A,D+,B)b_+c_+$	$=$	$R'U^2R^2UR'U'R'U^2FRF'$	(MBTC - 13)
$A_+B_-a_+b_+$	$=$	$B^2R^2F[B,R']F'R^2B^2[U,R]$	(MBT - 14)
$A_+B_+C_+(a,b,c)$	$=$	$U^2LUL'ULU^2L'$	(BCS & KO - 8)
$A_+B_+C_+(a,b,d)$	$=$	$U^2B'U^2BUB'UB$	(K. Fried - 8; p 24)
$A_+B_+C_+a_+b_+$	$=$	$B'U'B^2L'B'L^2U'L'U^2$	(D. Benson - 9; p 24)
$A_+B_+C_-D_-(a,b)(c,d)$	$=$	$F[U,R]F'\cdot B[L,U]B'$	(MBT - 12)
$A_+B_+C_-D_-(a,c+,b)$	$=$	$F[U,R]^2F'$	(DBS - 10)
$A_+B_+C_-D_-(a,c+,d+)$	$=$	$B[L,U]^2B'$	(DBS - 10)
$A_+B_-C_+D_-(a,b,c)$	$=$	$([F,U][U^2,F])^2$	(J. Trapp - 11)

D. SOME NON-U PROCESSES.

$(UF,DF)(UR,DR)$	$=$	$(F^2R^2)^3$	(DBS - 6; p 13)
$(UF,DF)(UB,DB)$	$=$	$F^2R^2_sB^2R^2_s$	(DBS & WK - 6; 17-B, pp 14,21)
$(UF,DF)_+(DR,UR)_+$	$=$	$FUF^2U'R'F'U'R^2UR$	(MBTC - 10; p 35)
$(UF,DB)(UB,DF)$	$=$	$R_sU^2_sL_sF^2$	(DBS - 8)

(See also various flips in Pretty Patterns.)

(BU,FU,FD)	$=$	$F^2R_sU^2R'_s$	(T. Varga & K. Fried - 6; p 23)
(FU,UR,RF)	$=$	$RU'R'F'L'B'U'BLFRU^2R'$	(GK - 13; 17-A-v, pp 14,21,28)
(FD,UL,RB)	$=$	$U^2DFDBR'B'D'F'U'RU'$	(GK - 13; p 28)
		a	

(The last two are rotations about the FUR-BLD axis.)

$UR_+DR_+UL_+DL_+$	$=$	$F^2_sLF^2U_sR^2BR^2_sF'L^2U_sB^2R'$	(MBT - 16; p 35, slice 4-flip)
$(FLU,FUR)_+(FRD,DRB)_-(FU,FR,DR)$	$=$	$FRF'R' = [F,R]$	(DBS - 4; p 15)
$(URF,ULB)(DLF,DRB)$	$=$	$(R^2U^2F^2)^6$	(A. J. Adamyck & WK - 18)
(FRD,UFL,RUB)	$=$	$U'LFDF'U'FD'F'UL'U = U'L[[F,D],U']L'U$	(DBS - 12; p 28)

(This last is also a rotation about FUR-BLD axis.)

(ULB,DLF,DFR)	$=$	$UR^2U'L^2UR^2U'L^2 = [[U,R^2],L^2]$	(DBS - 8, p 35)
$FLU_+FUR_+FRD_-DRB_-(FU,DR,FR)$	$=$	$[F,R]^2$	(DBS - 8, p 15)
$ULB_+UBR_-URF_+UFL_-(FL,BU,LU)(FR,FU,RU)$	$=$	$U'F^2U^2F'U^2F^2$	(? - 6)
$ULB_+UBR_-URF_+UFL_-(FL,FU,LU)(FR,BU,RU)$	$=$	$UB^2D^2FD^2B^2$	(MBT - 6)
$ULB_+UBR_-URF_-UFL_-DBL_-DRB_+DFR_+DLF_+$	$=$	$(LR^2F^2B')^4$	(MBT - 16)

Chris Rowley asks for an easy 11-cycle on edges and an easy 7 cycle on corners.

E. EDGE(S) INTO MIDDLE SLICE PROCESSES. (See also p 25.)

$UF \rightarrow FR \rightarrow LU$	by	$F'U^2L'ULU^2F$	(D. E. Taylor - 7)
$LU \rightarrow FR \rightarrow UB$	by	$FR'F'RF'U'F$	(A. H. Frey, Jr. - 7)
$LU \rightarrow FR \rightarrow UB$	by	$RU'R'U'F'UF$	(BCG - 7)
$FU \rightarrow FR \rightarrow RU$	by	$LF^2UFU'F^2L'$	(A. Kertesz - 7)

F. SUPERGROUP PROCESSES.

Here we list some processes which only turn face centres, which are indicated by F, etc. on the left of the following. See also pp 3, 18, 22, 34, 38.

U^2	$=$	$(UR_aU^2R'_a)^2$	(MBT - 12)
UF'	$=$	$F_sL_sU_s\cdot F'\cdot U'_sL'_sF'\cdot U$	(MBT - 14)
UD'	$=$	$R_sF^2R_s\cdot U\cdot R_sF^2R_s\cdot D'$	(D. Benson - 14; p 31)
	$=$	$R_aF^2R'_a\cdot U\cdot R_aF^2R_a\cdot D'$	(Z. Perjés - 14)

No simple UF is known to me.

G.   COMMENTS ON USEFUL NEW PARTIAL PROCESSES.

The following are some new partial processes with fairly simple structure which can be combined in various useful ways. The most important forms of combination are the following.

P is a process which acts on a face or slice in a minimal manner, such as the monoflip or monotwist. If P acts minimally on the U face, then $PU^nP'U^{-n}$ will only act on U.

P is a process which moves the pieces of one face to another. If P moves all the U pieces to D, then PD^nP' will only act on U.

In either case, if P has order 2, then $P^2 = I$, $P = P'$ and things are simpler. The reader is advised to study these processes carefully and note how they have been used in the above subsections as it is difficult to explain all the ways in which they might be used.

$R_a U^2 R'_a$ exchanges the RF and LB vertical columns and the adjacent edges UF, UB. This process has order 2.

$R^2B^2R^2U^2R_s$ moves all U pieces onto the B face, with two edges flipped.

$R^2F^2_sL^2$ moves all U pieces onto the D face, with two edges exchanged. This has order 2.

$FDF^2D^2F^2D'F'$ = M exchanges only (URF,**FLU**) = (C,D+) in the U face and has order 2. BCG call this a <u>Monoswap</u>. Perhaps it could be called a unicycle? This makes an easy way to get U corners in place once they are in an even permutation.

$BR'D^2RB'$ exchanges only (ULB,RFU) = (A,C+) in the U face and has order 2. Though it is a monoswap, its shape is not very useful. However $BR'D^2RB'U^2$ gives A_+C_- and a lot of 2-cycles. Hence its square yields A_-C_+. Gerzson Kéri says he had this from Rubik himself, so I call it <u>Rubik's Duotwist</u>.

$L'FD^2LF^2D'F$ only moves (A,C-)D- in the U face. This allows you to construct a number of twisting moves. Being due to Thistlethwaite I call this <u>Thistlethwaite's Tritwist</u>, though this is a thongue thwister!

5.9.   PRETTY PATTERNS.

A.   SYMMETRIES OF THE CUBE AND THE CENTRES GROUP.

Since many pretty patterns are based on symmetries of the cube, let us first examine the symmetries. There are three kinds of symmetry axis of a cube, exemplified by FU-BD, FUR-BLD and UD, with orders 2, 3 and 4 respectively. We can call these an edge axis, a corner axis (or main diagonal) and a face axis. The cube pieces fall into various orbits under these rotations and it is easy to see precisely which orbits or combinations of orbits can be rotated by processes on the cube. We already know how to carry out all the rotations about a corner axis. A corner axis rotation has six 3-orbits and two 1-orbits. Rotation about a face axis preserves layers and hence it is easy to obtain any achievable rotations. A 90° rotation has five 4-orbits, so it is not possible to rotate all the orbits together. A 180° rotation has ten 2-orbits and can be achieved, yielding the 4-spot pattern. Rotations about an edge axis have been less studied but the most interesting cases are given below. There are nine 2-orbits and two 1-orbits, so one cannot rotate all the orbits at once.

Reflections have been less studied, mostly because neither any corner nor the configuration of centres can be reflected. So any reflectional pattern must be viewed as reflecting just edges. Reflection with respect to the plane of a slice has four 2-orbits and four 1-orbits and can be readily carried out to yield a 2-X pattern. Reflection with respect to the plane through the FR and BL edges has five 2-orbits and two 1-orbits, so cannot be done in entirety. Reflection with respect to the plane of a central hexagon (i.e. the plane which perpendicularly bisects a main diagonal) has three 2-orbits, so again cannot be accomplished in entirety. Reflection or inversion with respect to the central point of the cube involves 6 2-orbits. This can be achieved by $F_s^2U_s^2R_s^2$.

In all the cases where a rotation involves an odd number of 2-orbits, it is not possible to rotate all the pieces as a unit with respect to the centres. This leads us to see precisely how the centres can be considered as moving with respect to the rest of the cube. The centres, being held together physically, can only be moved like the direct symmetries of the cube (i.e. reflections are not permitted). There are 24 of these (pp 3, 18). These are actually equivalent to S_4 which can be seen by considering S_4 as acting on the four main diagonals of the cube. However, we have only found 12 movements of the centres: I, eight 6-spots (two about each main diagonal) and three 4-spots (one about each face axis). The other 12 direct symmetries of the cube are odd permutations of the six centres and correspond to the rotations by 90° about face axes or 180° about edge axes. From the first paragraph, we know such rotations of all the pieces with respect to the centres are impossible. So the centres can only be moved with respect to the rest of the cube (or vice versa) by even direct symmetries of the faces of the cube. We call this the <u>centres group</u>.

We can view movement of the centres as resulting from physical movement of the cube. It is sometimes easier to view a pattern with respect to the main body of pieces and think of the centres as having been moved. Examination of all the symmetries of the cube shows that they are all even permutations considered on the set of all 26 pieces - 8 corners, 12 edges and 6 centres. Indeed they are all even on the set of 8 corners and on the set of 18 non-corners. Since all the achievable patterns on the cube, considering the centres as fixed, are even on the same 26 pieces, we have that all achievable positions of the 26 pieces are even permutations. For example, the 6-spot pattern can be viewed as six three cycles and two twists on the corners and edges or as two 3-cycles of centres. In the latter case, we must bring the cube back to having all the non-centre pieces in their original places by using a symmetry of the cube. That is, the two viewpoints differ by a symmetry of the cube, which is an even permutation. In particular, we can obtain only even permutations of the centres when everything else is fixed.

B.   THE SIMPLER PRETTY PATTERNS.

Here we summarize and give standard names to the simpler and better known of the pretty patterns.   Note: These names are based on the facial appearance. There are patterns which have the same facial appearances but a different relationship of faces. For example, there are at least two other examples of a 6-X pattern than given here. (pp 28, 30).

2-X	$= (F^2R^2B^2L^2)^3$	
	$= (F^2R^2)^3(B^2L^2)^3$	(WK - 12) (See p 53 for an 8-move.)
4+	$= (F_aR_a)^3(R_aF_a)^3$	(DBS - 24; pp 11,21) (See p 48 for a 10-move.)
4-X	$=$ two 2-X patterns along different axes or 4+ & 4-spot	
(Slice) 4-flip	$= F_s^2LF_s^2U\ R_s^2BR_s^2F'L^2U_s'B^2R'$	(MBT - 16; p 31,35,36,45)
4-spot	$= R_s^2U\ F_s^2U_s$	(K. Fried & WK - 8; pp 11,20,31
4-Z	$= (F_aR_a)^3U_a^2$	(DBS - 14; pp 11,21)
6-2L	$= F_aU_aR'_a\ F_a$	(DBS - 8; pp 11,21)
6-bar	is not achievable	(DBS; p 28)

6-spot $= R_s F_s U_s R_s$ (WK - 8; pp 11,20)

6-X $= R_s^2 F_s^2 U_s^2$ (WK - 6; pp 11,20)

8-flip $= (R_a F_a U_a)^2$ (MBT - 12; p 35)

12-flip   = 8 flip and (slice) 4-flip   (MBT - 26; pp 28,31,35)

Tricolours = $C(R,F,D)B^2 C(R,F,D)^{-1} C(L,B,D)F^2 C(L,B,D)^{-1} \cdot F^2 R_s^2 B^2 R$

where $C(R,F,D) = R^2 FDR^2 D'R$ (D. Benson - 31; p 31)

U-4-flip $= R^2 B^2 R^2 U^2 R_s \cdot B \cdot R_s' U^2 R^2 B^2 R^2 \cdot U$ (MBT - 14; pp 44)

Zig-Zag $= (F_a R_a)^3$ (WK - 12; pp 11,21)

C. MORE COMPLEX PRETTY PATTERNS.

Richard Walker has extensively studied these patterns and found improvements to most of them.

The <u>4-bar</u> pattern on page 28 can be viewed as rotating the 4 vertical corner columns with respect to the rest by 180°. (The phrase 'with respect to the rest' will henceforth be omitted and understood.) The pattern is achieved by

$F_s^2 R^2 B^2 R^2 F_s^2 L^2$ (DBS - 8; p 28),

$R_a U^2 R^2 U^2 R_a$ (N. J. Hammond - 8),

$(R^2 F^2 L^2)^2$ (R. Walker - 6).

Replacing the last L^2 by R^2 in the first and third of these gives <u>Crossbars</u>, as does replacing the last R_a by R_a' in the second of these.

My <u>Double-cube</u> pattern described in the first paragraph of section 10-E (<u>pp 28, 29</u>) can be achieved more easily by moving the rest of the cube instead. Consider

(RDF,UBR,FLU)(RD,UB,FL) = BL'D²LDF'D²FD'B'   (R. Walker - 10).

This acts on three corner-edge pairs such as RDG, RD based at the three corners closest to URF. It rotates them 120° anti-clockwise about the FUR-BLD axis. Combining this with an appropriate reflection of itself, namely

(RBD,ULB,FDL)(RB,UL,FD) = F'RU²R'U'BU²B'UF

yields the double cube in 20 moves.

Walker also gives

$(FUR)+(FU,UR,RF)(FRD)\_(BUL)+(BU,UL,LF)(BLD)\_ = R_a UR_a' F_a' UF_a$ (R. Walker - 10),

which is equivalent to two of Ahrens' processes A of page 29. Using this once and two of Ahrens' processes yields the rotation of four non-adjacent corners and their adjoining edges in 34 moves instead of 48.

$U(R_s^2 F_s^2)^2 U'(R_s^2 F_s^2)^2$ (R. Walker - 14)   rotates the corners 90° clock-wise about the UD axis. (Directions of rotation are given with respect to looking at the first named end of the axis. E.g. here I am looking at the U face to see the 90° clockwise rotation.) This is a 4+ pattern.   (pp 33,35)

$F_s^2 R^2 UF^2 R^2 D$ (R. Walker - 10)   rotates the U corners as though U had been applied and the D corners as though D had been applied.

$F_a^2 U_a F_a^2 R^2 U_a$ (F. O'Hara - 12; pp 11,21,33,35)

or $F_s^2 R^2 U^2 F_s^2 R^2 D^2$ (R. Walker - 10; pp 11,21,33,35)

rotates the corners 180° about the UD axis. This is the usual 4+ pattern. Conjugating the latter by R, and noting that slice-squares commute, we get an 11 move process which rotates the corners 180° about the FU-BD axis as first given by Ahrens on page 31. This is a 6+ pattern.

Consider

W(R,U,F) = RUF²U'F'R'U'R²UF = $(FL,RD)_+(FD,RB)_+$ (R. Walker - 10). Then

W(R,U,F)·W(F,R,U) = (FL,RD,UB)(FD,RB,UL)   (R. Walker - 19; p 30)

is a 120° anticlockwise rotation of the <u>central hexagon</u> about the FUR-BLD axis.

$$U'R^2_{a}F_{a}U^2F'R'UB'RU_{s}F'RU'FR'U_{s}'R^2F^2U = FL_+(FD,LU)(RD,BU)RB_+ \quad (Walker-22)$$

is a rotation of the same central hexagon about the FL-UB axis.

There are two belt patterns on pages 30 and 31.

$$RUF^2D'R_{s}F_{s}D'F'R'F^2\overline{RU^2FR}^2F'R'U'F'U^2FR \qquad (R. Walker - 23)$$

obtains the first one which Walker names the Worm.

$$BR_{s}D'R^2DR'B'R^2 \cdot UB^2U'D\mathbf{R^2}D' \qquad (R. \overline{Walker} - 16)$$

obtains the second one, which Walker names the Snake. Modifying the last to

$$BR_{s}D'R^2DR'B'R^2 \cdot UB^2U'DB^2R_{a}U^2R'B^2D' \qquad (R. \overline{Walker} - 22; p\ 31)$$

gives the pattern of Ahrens on page 31 in which the 12 edges and **centres** adjacent to RUB and LFD are rotated 180° about the FR-BL axis. This is perhaps the best example where it is easier to describe the pattern as if the centres had moved and they have been moved in an odd permutation.

The corners can be rotated 120° anticlockwise about the RUB-LFD axis, as discussed on pages 28 and 30, by using the 11 move 180° rotation about a face axis, applying it two times as follows:

$$R'L^2F^2_{s}U^2R^2_{s}F^2_{s}D^2R' \cdot D'U^2F^2_{s}R^2_{s}U^2F^2L^2D' \qquad (R. Walker - 22).$$

This is a 6+ pattern. We can then rotate the centres as well to get the 6-X pattern of page 28 and 30. The result can also be viewed as rotating the edges and the following accomplishes the result directly.

$$B'R'U'F_{s}U^2R^2_{s}F^2_{s}D^2R \cdot DU^2F^2_{s}R^2U^2F^2_{s}L^2D \qquad (R. Walker - 25)$$

Oliver Pretzel sends a 6-U pattern, which is like the worm but with alternate edges missing.

$$L'R^2 \cdot F'L'B'UBLFRU'R' \cdot R^2L = (FL,UB,RD) \quad (13\ moves)\ followed\ by\ the\ 6\text{-}spot:$$

$R_{s}F_{s}U_{s}R_{s}$ gives the pattern in 18 moves. There is also a 4-U:

$$R_{a}U^2R' \cdot F_{a}U^2F' \qquad (DBS - 10).$$

5.10.   THEORETICAL DEVELOPMENTS AND PROBLEMS.

A.   THE U GROUP.

This is the group of U processes. Analysis as in the general case shows there are $\dfrac{4!\ 4!}{2}\dfrac{3^4}{3}\dfrac{2^4}{2} = 62208 = 2^8 3^5$ possible patterns on one face, as previously on page 30. The shortest non-trivial U processes seem to be of the form B[L,U]B' (6 moves). The tables in BCS show that any pattern is achievable in at most 31 moves. The most difficult patterns in their lists are: (A,B)(b,d) (with any orientation); $A_+B_-a_+b_+c_+d_+$; $A_+B_-C_+D_-$ requiring 13, 18 and 18 moves respectively. Gerzson Kéri conjectures that any U process is achievable in at most 18 moves.

B.   GENERATION OF THE WHOLE GROUP.

Frank Barnes has shown that the group of the cube is generated by just two elements (page 32). Marston D. E. Conder has shown that it can be generated by two elements γ, δ of orders 2 and 4 and that orders 2 and 3 will not suffice. It is well known that two elements of order 2 can only generate the symmetries of a regular n-gon, where n is the order of $\gamma\delta$. (When n = 1 or 2, the n-gon is a bit degenerate.) Elements of orders 3 and 3 can only generate even permutations on corners and on edges, hence cannot generate the whole group. It is not clear if Conder's generators also generate the supergroup.

Explicitly, Conder shows that the following serve:

γ = (UR,RB)(DR,UB)(FR,FL)(UL,LB)(RFU,RUB)(DFR,DBL)(UFL,BUL)

  = (UR,BR,DR,FR)(UF,UB,DF,DB)(UL,FL,DL,BL)(RFU,DFR,LFD,UFL)(RUB,BDR)(DBL,ULB).

He gives 92 and 87 move processes for these results, noting that one could do them in many fewer moves. In a revised version of his note, he uses

γ' = (FL,LB)(BR,FR)(DL,UB)(DR,UF)(DF,DB)(RFU,RUB)(DFR,DBL)(UFL,BUL)

and he shows that $\gamma'\delta$ has the maximal possible order of 1260 (pp 18,22,27).

C. PRESENTATIONS.

Krystyna Dałek asks about presentations for various cube groups. A presentation is a definition of a group by giving two sets: a set of elements, called generators, and a set of identities (which group theorist call relations or relators). The group defined is the smallest group containing the elements and satisfying the identities. For example, the group of symmetries of the square (pp 4-10, 18, 19) is defined by $<R,V|R^4 = V^2 = I, VR = R'V>$ or by $<V,D_1|V^2 = D_1^2 = (VD_1)^4 = I>$. (The vertical bar can be read as 'such that'.) Normally one wants the sets to be irredundant and/or minimal in some sense but this is not a clear idea nor an easy task. Except for the simplest cases, such as the slice-squared group, I do not know of any presentations for cube groups.

D. ORDERS OF ELEMENTS.

J. B. Butler has sent the following example of an element of the maximal order 1260.
$$RF^2B'UB' = (DFR,FDL,LUF)\_(URF,BLD,DRB,UBR,BUL)_+(FU,FD,LU,BR,DR,FL,FR)_+\cdot$$
$$(LB,UR)_+(UB,DB)$$
He asks if this is the minimal number of moves or the minimal length for an element of order 1260. By the length of a process, we mean the total number of $90°$ turns in the process. Thus R and R' count as length 1 but R^2 counts as length 2. For example, Butler's element has length 6. The effects of a process on both corners and edges is even if and only if its length is even. Since any process of order 1260 seems to have an even effect on corners, it must have even length. Since processes of length 2 cannot affect enough pieces, Butler's question can be resolved by checking the processes of length 4.

Butler has a program to determine orders of processes. He finds that about one process in a thousand has order 1260, but that elements of order 3 are scarce. This led me to consider the following.

Problem. What are the possible orders of elements in the group of the cube?

To attack this problem, one needs to know a corollary of Lagrange's Theorem (the first basic result of group theory): The order of any element must divide the order of the group (i.e. the number of elements in the group). In our case, the order is $N = 2^{27}3^{14}5^37^211$. Hence we can have no element of order 13, 17, 19, ... It is easy to find elements for each order less than 13, so 13 is the first nonorder. However, this is such an obvious reason that we will call a nonorder trivial if it does not divide N. The first nontrivial nonorder is 25, though this is still fairly obvious since an element of order 25 must contain a 25-cycle! A nonorder is trivial exactly when it is divisible by a prime greater than 12. We say that a nonorder is obvious if it is divisible by a prime power greater than 24 (which is the longest possible twisted cycle length). The obvious cases cover most of the cases. The first nonobvious case appears to be $385 = 5\cdot7\cdot11$ which cannot be an order since we cannot have 5, 7 and 11 cycles simultaneously.

Problem. How many elements are there of each possible order?

It is feasible, though tedious, to actually compute the number of elements of each order. We sketch the method for the case of m = 3, which is fairly simple. If P has order 3, then it can only contain 3-cycles, 1-cycles and twisted corner 1-cycles. The number of elements of orders 3 or 1 can be separately computed for corners and for edges and the product of the results will be the total number of elements of orders 3 or 1. Since only I has order 1, we merely subtract 1 from the product to obtain the number of elements of order 3.

Consider the case of a 1-cycles and b 3-cycles of corners. Note that $a + 3b = 8$. For $a = b = 2$, the typical form of P is
$$P = (C1)(C2)(C3,C4,C5)(C6,C7,C8).$$

The 8 corners C1, C2, ..., C8 can be written into the typical form in 8! different orders (we are ignoring orientations for the moment). But the b 3-cycles can be permuted among themselves in b! different ways for the same P. Similarly the a 1-cycles can be permuted in a! ways. Further, a 3-cycle can be cycled in 3 ways without changing P. That is, $(C3,C4,C5) = (C4,C5,C3) = (C5,C3,C4)$. Taking all b 3-cycles into account gives 3^b = 9 ways of cycling without changing P. If we ignore orientations, we get $\dfrac{8!}{a!1^a b!3^b}$ permutations P with a 1-cycles and b 3-cycles and it is easy to see how this generalizes to arbitrary cycle structures in S_n.

Now we must account for orientations. Each of the 8 corners has 3 orientations, which would seem to give a factor of 3^8. But consider a 3-cycle. If we twist all three pieces the same way, it is the same 3-cycle, e.g. (ULB,UBR,URF) = (LBU,BRU,RFU) = (BUL,RUB,FUR). Thus we only have 3^2 different forms for each 3-cycle, giving 9^b for all b 3-cycles. For 1-cycles, all the orientations are the same. But the 1-cycles may be twisted, provided the total twist is 0 (mod 3). In order to avoid getting confused with the a! ways to permute the objects in the 1-cycles, we associate the twisting with the objects, rather than the positions within the typical form. (That is, we want our permutations of the 1-cycles to take $C1_+C2_-$ to $C2_-C1_+$ rather than $C2_+C1_-$, say.) There are 3^a ways to associate twists with the a corners in the 1-cycles and precisely 1/3 of these will total to 0 (mod 3). Thus we have 3^{a-1} ways to twist a 1-cycles. Putting this all together, we have

$$\frac{8!3^{a-1}9^b}{a!1^a b!3^b}$$ elements in the cube group with a 1-cycles and b 3-cycles.

Computing for (a,b) = (8,0), (5,1), (2,2) and adding yields 2187 + 81648 + 272160 = 3 55975 corner elements of orders 3 or 1. Similar reasoning on edges yields the same expression with $8!3^{a-1}9^b$ replaced by $12!4^b$ and calculation for (a,b) = (12,0), (9,1), (6,2), (3,3), (0,4) yields 1 + 1760 + 591360 + 31539200 + 63078400 = 952 10721. Multiplying the two results together and subtracting 1 gives us 2245 55950 08000 $\simeq$ 2.25 x 10^{13} elements of order 3. Dividing this by N $\simeq$ 4.33 × 10^{19} gives a probability of about 5.17 x 10^{-7} of finding an element of order 3 at random. That is, there is about half a chance in a million of doing so. So Butler's observation seems pretty much spot on.

For composite orders, the analysis is a bit more involved. First, the kinds of cycles occurring are more complex and more numerous. Second, one obtains all the elements whose orders divide the order and these must be subtracted in a more complicated manner.

E. THE TWO-DIMENSIONAL PROBLEM.

Nicholas J. Hammond asked me about the 4 × 4 'cube'. Consequently I have examined the 2 × 2, 3 × 3 and 4 × 4 cases. There are two conceptual decisions one must make. I. Are the squares two-sided or not? That is, if we have a pattern: 1 2 and we invert the bottom row,
4 3
do we recognise that the 3 and 4 are now turned over? On the 2 × 2, it is easy to deal with this - a number is upside down if and only if it is in a position of the opposite parity. On the 3 × 3, the edge pieces remain in place and can be flipped over. II. On the 4 × 4 and larger squares, can we turn over all rows and columns or just the edge ones? In the latter case, the square is equivalent to the 3 × 3 two-sided square. Hence we will assume the first case, i.e. that all rows and columns can be turned.

In the 2 × 2 one-sided case, we get all 24 permutations of the 4 corners, but since there is no fixed orientation, we can divide by

4 by rotating the square, or by 8 if we can also turn the whole thing over which is equivalent to allowing reflections. In the two-sided case, we get 24 patterns, but rotation by 90° yields another 24. The symmetry allows us to divide by either 4 or 8 as before.

In the 3 × 3 one-sided case, the Greek cross or + pattern of edges and the centre can be used to define the orientation of the whole square. Then the corners can be put into all 24 permutations with respect to this orientation and there is no symmetry factor to divide by. In the two-sided case, the corners give 24 permutations as above and the edges can be flipped in pairs. Since the number of edge flips has the same parity as the permutation of corners, we get $4!2^4/2 = 192$ patterns and this group is the same as $<F^2,R^2,B^2,L^2>$ on the cube. (Group theorists say the groups are <u>isomorphic</u>.) This group can also be considered on the Magic Domino and the problem can be simulated by a domino with a pair of U-D pieces clipped together so that the turns of the U and D faces cannot be done.

For the 4 × 4 case, both the one-sided and the two-sided problem appear to have the same group. There are four orbits of order 4, but each move is an even permutation of the 16 squares. All $(4!)^4/2$ = 1 65888 positions can be achieved but there are 4 possible symmetries, so there are only 41472 distinct patterns.

For larger problems, I cannot yet see if all the expected patterns can be achieved or if there is some reason to expect fewer patterns. Once this is resolved, one could attack the m × n problem.

F.   CAYLEY GRAPHS AND ANTIPODES.

For any group given in terms of generators, say A, B, ..., the <u>Cayley graph</u> of the group is the diagram consisting of points corresponding to all the elements of the group with an arrow labelled A pointing from P to Q if Q = PA. We normally indicate the labelling by using different kinds of lines or arrows. Generators of order 2 can be denoted by two-way arrows. For example, if we consider the symmetries of the square (pp 4-10, 18,19) as generated by R and V, then the Cayley graph is shown in Figure 5.7.

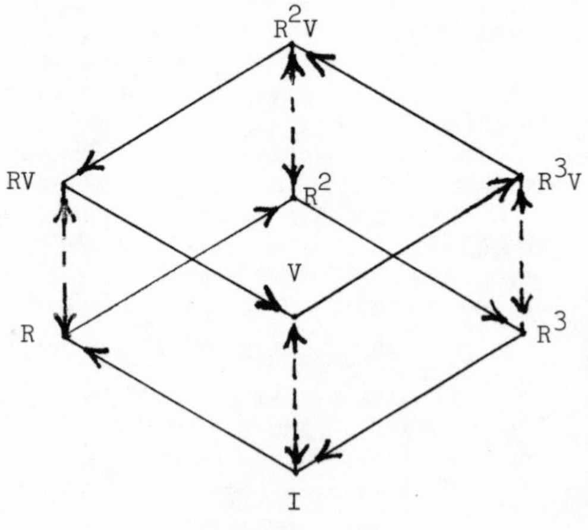

A path in a Cayley graph is a sequence of generators. E.g. there are two paths from I to RV: RV and VR'. (Note that going backward along an arrow means taking the inverse of the generator.) A closed path is an identity, e.g. RVRV = I.

In the Figure we see there is a unique <u>antipode</u> to I, i.e. a point at maximal distance 3 from I. This distance is called the <u>diameter</u> of the graph or the group.

FIGURE 5.7

Michael Holroyd has examined the Cayley graphs of the slice-squared group, $<F^2,R^2,B^2>$, $<F^2,R^2,B^2,L^2>$ and $<F_s,R_s>$, where the last one is taken with respect to corner coordinates (p 20). In the last case and in general, the Cayley graph is neater if we do not consider F^2, etc. as generators. However this means that the distance in the graph will be the length of a process (p 50) rather than the number of moves. Hence the diameter **may** be greater than the maximum number of moves. The groups

considered by Holroyd have 8, 96, 192 and 192 elements respectively and each has a unique antipode: the 6-X pattern of distance 3;
(FU,FD)(BU,BD) = $F^2R^2F^2R^2F^2B^2R^2B^2R^2B^2$ of distance 10;
(FU,FD)(RU,RD)(BU,BD)(LU,LD) = $F^2L^2F^2L^2R^2F^2L^2B^2$ of distance 8 (a 2-X pattern);
and (UF,DB)(UR,DL)(UB,DF)(UL,DR) = $F_sR_sF_sR_s\cdot R_sF'_sR'_sF_s$ of distance 8 (in slice

moves). The last pattern has all the U and D edge reflected with respect to the centre of the cube. It has 2 X faces and 4 H faces (a = c = d $\neq$ b in Figure 13 on page 11) and is two slice-squared moves away from the 4-spot.

Holroyd wonders if the whole group of the cube has a unique antipode. Resolving this may require the complete description of God's algorithm (p 34). He suggests that either the 12-flip (pp 28,31,35,48) or the 12-flip combined with the ordinary 6-X pattern of the slice-squared group (pp 11,20,48) might be an antipode.

However, multiple antipodes can occur. The slice group has 32 antipodes at 5 moves from I. (See page 31.) The square group acting on corners, ignoring both edges and centres, which is the same as the square group on the 2 × 2 × 2 cube, has 5 antipodes at distance 4, as described in the next section.

G. THE SQUARE GROUP - 2 × 2 × 2 CASE.

On the 2 × 2 × 2 cube, square moves and antislice moves are the same. Hence the following analysis also is of interest in understanding the antislice group.

Since we have no orientation forced on us, we may fix some corner. Let us fix the BLD corner by agreeing that whatever piece is there is correct. Then we need only consider the actions which leave BLD fixed, i.e. $<F^2,R^2,U^2>$. The square group has two orbits of corners: the four corners at distance 0 or 2 from BLD and the four corners at distance 1 or 3. Here distance is measured as steps along the edges of the cube, so the first orbit consists of BLD, UFL, UBR, DFR and the four corners are at the edges of a regular tetrahedron. Once BLD is fixed, the first orbit has only three remaining elements and the other has four elements. Each square turn is two 2-cycles, one in each orbit. Square turns do not lead to changes of orientation - each U or D face is always in the U or D direction - so we need only consider positions.

After some playing, we decide that the action on the 4-orbit actually determines the action on the 3-orbit, so that there are only 24 possible patterns. Exercise. Can you see this?

To see this result, we first need to have a convenient labelling for the corners. Let us label them as in Figure 5.8. Then we have
F^2 = (A,B)(1,2), R^2 = (A,C)(2,3), U^2 = (A,D)(1,3).

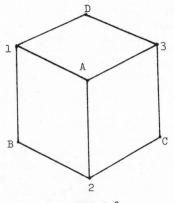

FIGURE 5.8

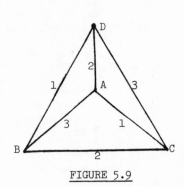

FIGURE 5.9

Now consider the tetrahedron with vertices at the lettered corners, as shown in Figure 5.9. Three of the tetrahedron edges correspond to the numbered corners and we number the three remaining edges with the numbers of their antipodal edges. E.g. the edge AD has the same number as the edge BC. Then we see that the three square moves can be considered as reflections of the tetrahedron with respect to the planes through two of the edges of the set B, C, D. That is, F^2 acts like (A,B)(1,2) which is a reflection of the tetrahedron with respect to the central plane through the vertices B and C. In this reflection, the edges labelled 3 are carried to themselves but the two edges labelled 2 are carried to the two edges labelled 1. Thus the cycle (1,2) simply describes what happens to the pairs of edges when the symmetry (A,B) of vertices is carried out. Stated another way, the action of the square group on vertices determines its action on edges. Translating this back to the cube, we say that the action on the 4-orbit determines the action on the 3-orbit. Hence we can get at most 24 patterns corresponding to the 24 permutations of the 4-orbit. It is easy to see that we can indeed get all these permutations.

The Cayley graph is awkward to draw but one finds 5 antipodes at distance 4, namely the three pairs of 2-cycles and the two 3-cycles not affecting A. There are four permutations of the 4-orbit which are the identity on the 3-cycle, namely I and the three pairs of 2-cycles. These permutations form what is called the <u>kernel</u> of the action of S_4 on the 3-orbit.

H.   THE ANTISLICE GROUP.

Thistlethwaite has written out his analysis of the antislice group, but we cannot find any way to simplify the details for presentation here. Instead we provide a more detailed outline of the structure than was given on page 35. (See also pp 20-21 for the idea of corner coordinates.) He proceeds by fixing one corner and then examining one slice. The edges of the slice can be permuted in all 24 ways. However one can then only permute the other two slices in 4 ways each (comprising I and the three pairs of 2-cycles). Thus there are $24 \cdot 4 \cdot 4 = 384$ edge positions. For a given permutation of the edges, one can only flip all the edges in 0 or 2 slices (the 8-flip, see pp 35, 48) giving 4 orientation patterns of edges. Then one can only achieve four motions of the corners and centres as produced by the zig-zag pattern (pp 11, 21, 48), which gives just the two pairs of 2-cycles on the corners not in the orbit of BLD and also on the centres. This gives us a total of $384 \cdot 4 \cdot 4 = 6144 = 2^{11}3$ positions. Frank O'Hara is trying to find the diameter of this group.

I.   THE SQUARE GROUP - 3 × 3 × 3 CASE.

In section 5.10-G, we have seen that the square group, considered only on corners and ignoring centres and edges, has 24 elements. We proceeded by fixing the BLD corner first. Now BLD has four possible positions in relation to the centres, so we have $24 \cdot 4 = 96$ positions of corners with respect to the centres. We can view the four possibilities as the ways of moving centres with respect to the corners and these turn out to be just the identity and the three 4-spots. (<u>Exercise.</u> Show the 4-spots are in the square group. Is there an easy way to see that the 6-spots are not in the square group?)

One can get the corners right with respect to the centres in at most four moves, despite the fact that it can take four moves ignoring the centres. The answer is to skilfully choose the corner in the previous approach. If both the corner orbits are in even permutations, put BLD in place. We now have an odd permutation of the 4-orbit and all these can be achieved in three moves. If either orbit has a corner correct with respect to the centres, take it as BLD and restore the rest in at most four moves. Now suppose the orbit of BLD is in an odd permutation and no corner of it is correct. Then the orbit must be in a 4-cycle.

Applying one of B^2, L^2, D^2 will put BLD correct with respect to the centres and leaves a 3-cycle. Now we can restore the 4-orbit in three moves unless it is one on the antipodal positions. These are the three pairs of 2-cycles and the the two 3-cycles leaving A fixed. But the pairs of 2-cycles correspond to the identity on the 3-orbit, which we do not have. Hence the 4-orbit must be in one of the two 3-cycles which leave A fixed. But since we started with a move that did not affect A, A must have been in its correct place and we could have used it as BLD.

Thistlethwaite says it is probably easier to do all the positions than try to dream up an argument like the preceding paragraph.

Now we consider the edges. The square group never changes orientations of edges. Further each move carries each slice of four edges into itself and each move is an even permutation of edges comprising 2-cycles in two different slices. Thus there can be at most $(4!)^3/2$ $= 6912 = 2^8 3^3$ edge patterns. Now $(F^2R^2U^2R^2)^2 = (FD,FU,BU)$ (MBT - 8) which allows us to get any even permutation in one slice and $(F^2R^2)^3 = (FD,FU)(RU,RD)$ which allows us to make any two slices odd, with both processes leaving everything else fixed. Hence all 6912 edge positions can be achieved, leaving corners and centres fixed and the square group has $96 \cdot 6912 = 6\ 63552 = 2^{13}3^4$ elements as given on page 26.

Thistlethwaite has found that 6900 edge patterns can be obtained in at most 12 moves and the remaining 12 can be obtained in at most 14, but he hasn't shown that they can't be done in less than 14. By combination of corner and edge processes, he has shown that every position in the square group is achievable in at most 17 square moves and some positions require 15. It is possible that these numbers could be reduced by use of non-square moves.

J. THE TWO GENERATOR GROUP.

Alexander H. Frey, Jr., has asked about finding a restoring algorithm for <F,R>. That is, if someone confuses your cube using only F and R moves, how do we restore it using only F and R moves? Conceptually this is the same as understanding what patterns are in the group, since we can only be sure that we understand the group if we can see how to achieve every pattern in it. Frey asks the same problem for other groups, but now we consider just <F,R>. Using results from Frey, BCG, Benson, D. E. Taylor and Thistlethwaite, I have a nice analysis of this group and it turns out to contain a remarkable and unique phenomenon of group theory.

First we observe that edge orientations do not change - a F or B face of a piece is always F or B or U or D, while a R or L face is always R or L or U or D. Otherwise stated, FR can only go to FU, FL, FD, UR, BR, DR. Now it is not too hard to see that we can move any four edges (of the seven affected by F and R) onto the R face. Then F^2R^2F puts the four edges of the R face into UF, DF, UR, DR. Then $P_1 = (F^2R^2)^3$ gives just two 2-cycles of edges and the pieces can be conjugated back to wherever they came from. Hence we can achieve and even permutation of the seven edges without affecting corners. Since F or R is odd on edges, we can get all of S_7 acting on the seven edges, though we are then affecting corners.

Now we consider the action on the corners. We might expect to obtain all 6! permutations of the 6 corners which the R and F faces contain. We shall see that we only get 5! Number the six corners as shown in Figure 5.10. Then $F = (1,2,3,4)$ and $R' = (3,4,5,6)$. We want to determine what subgroup of S_6 is generated by these two 4-cycles.

Consider the 6·5/2 = 15 distinct
pairs of numbers ij, $1 \le i < j \le 6$.
These can be viewed as the diagonals
(and sides) of a hexagon, forming a
graph known as the <u>complete graph</u>
on six vertices, K_6, and one can
imagine these pairs as the edges of
the 5-simplex. (The 2-simplex is an
equilateral triangle, the 3-simplex is
a regular tetrahedron, ...) We form
five triples of these edges as follows.
A = {12, 35, 46}
B = {16, 23, 45}
C = {15, 26, 34}
D = {14, 25, 36}
E = {13, 24, 56}

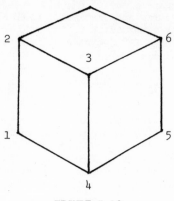

FIGURE 5.10

Now if we examine carefully the behaviour
of F and R, we see that they permute our
five triples among themselves as follows:
F = (A,B,C,D), R = (B,C,D,E). E.g.
F carries 12 to 23, 35 to 45, 46 to 16,
so F carries A to B. Group theorists say that we have induced an action
of <F,R> on the set {A,B,C,D,E}. We can now see why we only get 5!
positions of corners in our group, but there are two details to check.

We must verify that different elements of <F,R> give different
actions on the set {A,B,C,D,E}. Because of the group properties, we have
only to check that the identity action on our edge triples only arises
from the identity in <F,R>. Suppose we have some permutation P of
{1,2,3,4,5,6} which gives rise to the identity on {A,B,C,D,E}. Suppose
P(1) = 2. Then 12 must be carried to 12 and we must have P(2) = 1. Then
B being fixed says that 16 must be carried to 23 and we must have P(6) = 3.
Likewise 23 must be carried to 16 and P(3) = 6. But now C being fixed
requires 15 to be carried to 26 so P(5) = 6, contrary to P(3) = 6. Similar
reasoning shows that P(1) cannot be 2, 3, 4, 5, 6, so P(1) = 1 is the only
case left. Then the same sort of reasoning shows P(2) = 2, ..., P(6) = 6.
That is, the only way P can induce the identity on our edge triples is if
it is already the identity on corners.

Thus our mapping of actions is a one-to-one correspondence between
<F,R> and <(A,B,C,D),(B,C,D,E)>. Group theorists call this an isomorphism
since it is also a correspondence of the operations of the two groups.

The other detail to check is that we can achieve all of S_5. One
can work in either of the two groups. The action on the 5 edge triples
is easier to calculate but less easy to visualize, so we will work on
the set of 6 corners of the cube. The commutator [F,R] gives (2,3)(4,5)
when considered just on the corners, and its cube actually does this action
without affecting anything else. Using conjugates of [F,R]³, we can
easily get corners 1 and 2 into their correct places without affecting
edges. This gives already 30 cases. Now the only permutations of {3,4,5,6}
which act correctly on our edge triples are those generated by R. Of these,
only I and R^2 are even. Then I or $F'R^2[F,R]^3R^2F$ allows us to obtain the
effect without affecting edges. (If we don't care about edges, then I or
R^2 suffices and all of the argument in this paragraph can be similarly
simplified.) Since F or R is odd on corners, we can get all of S_5, though
we must affect edges for the odd permutations.

[Since every element of S_5 has a corresponding element of S_6, group
theorists say we have a <u>representation</u> of S_5 in S_6. This representation
is called <u>faithful</u> because the correspondence is one-to-one. This
representation is a famous and unique phenomenon in group theory and
worth describing a little bit further. S_5 has 24 5-cycles which fall into
6 cyclic subgroups of order 5. The action of S_5 on these 6 subgroups

is the action induced by conjugation of elements. It turns out that this action is identical to the action of <F,R> on our six corners. A correspondence between corners and subgroups of S_5 considered on {A,B,C,D,E} is given by:

1 → <(C,D,A,B,E)>
2 → <(D,A,B,C,E)>
3 → <(A,B,C,D,E)>
4 → <(B,C,D,A,E)>   =   <(A,E,B,C,D)>
5 → <(A,D,E,B,C)>
6 → <(A,C,D,E,B)>

The action is identical to the group of collineations of the projective line over the field Z_5 of five elements and to the group PGL(2,5) of 2×2 nonsingular matrices over Z_5 with scalar multiples identified. (See BCG, where this phenomenon is studied as the Tricky Six Puzzle and see also pp 107-109 of "Groups and Geometry" described in the Bibliography.)]

Returning to the group <F,R>, we need to analyse the orientation changes which corners can enjoy. Now $A = (R'FRF'R'F)^3$ leaves edges fixed and only twists corners as $1_-2_-3_+4_-5_+6_+$ (using an obvious notation). Then $AF'A'F = 3_-4_+$ and we can use this to obtain any corner twists that we are allowed, and without moving edges. (If we don't care about edges, things are much easier). All together then, the two generator group has $7!5!/2 \cdot 3^6/3 = 734\,83200 = 2^6 3^8 5^2 7$ elements as given on page 26.

K.   THE GROUP <F,R²>.

Having analysed the groups <F,R> and <F²,R²> (see p 20 for the latter), I decided to examine <F,R²>. First note that neither edge nor corner orientations ever change. (Edges are as in the previous subsection, but even easier. The F face of a corner is always F or B.) Clearly RU and RD are always either fixed or exchanged. If F^1 puts xF into RF, then $F^1R^2F(F^2R^2)^3F'R^2F^{-1} = (RU,RD)(RB,xF)$. Hence we can obtain any permutation of the 5 edges {FD, FL, FU, FR, BR} and of {RU, RD} such that the actions on the two sets have the same parity, i.e. the permutation is even overall, and these are obtainable without moving corners. Using F, which is odd on edges, we can get all possible 5!2! permutations, but then we move corners.

For corners, we proceed as in our discussion of F,R , but we shall not try to find corner processes leaving edges fixed. We observe that $FR^2F'R^2 = [F,R^2] = (2,3,4,6,5)$ is a 5-cycle on corners (we are ignoring the edges). If corner 1 is not in place, apply the 5-cycle until it is at position 3 and then apply F^2. We then use the 5-cycle to get corner 2 in place. Now in <F,R>, we know the only even permutations leaving 1 and 2 fixed are I and R^2, and these are achievable by applying I or R^2. Hence can achieve all the even permutations of S_5. Since F is odd on corners, we also get the odd ones. Hence <F,R> and <F,R²> have the same effects or corner positions. Thus $|F,R^2| = 5!5!2!/2 = 14400 = 2^6 3^2 5^2$

L.   CONJUGATION AND CYCLE STRUCTURE.

Recall that the conjugate of a permutation is a permutation with the same cycle structure. (See p 13.) We need to examine this in more detail . Suppose P carries A to B. In normal mathematical notation, we write this as P(A) = B (as on pp 5-6). Now PQ(A) normally means P(Q(A)), i.e. first apply Q and then P. But we have defined PQ to mean that P is applied first, so we must read PQ(A) as Q(P(A)) as mentioned on page 9.

Now P(A) = B means that (...,A,B,...) occurs somewhere in the cycle representation of P. Consider now $Q^{-1}PQ$ applied to Q(A). This is $Q(P(Q^{-1}(Q(A)))) = Q(P(A)) = Q(B)$. Thus we will have (...,Q(A),Q(B),...) in the cycle structure of $Q^{-1}PQ$ and the cycle representation of $Q^{-1}PQ$

is the same as that for P with each entry A replaced by Q(A). For this reason, group theorists usually call $Q^{-1}PQ$ the conjugate of P by Q. In our usage, it was easier to **have** QPQ^{-1} as then Q is the process which moves the pieces into the positions on which P acts.

We now wish to consider the inverse problem: if two permutations have the same cycle structure, are they conjugates? (Cycle structure can be formally defined as the number of 1-cycles, the number of 2-cycles, ..., so two permutations have the same cycle structure if they have the same number of cycles of each given length.)

The answer to the inverse problem is certainly positive when the group is S_n - simply write down the cycle representations of both permutations with the cycle length increasing and then let Q be the correspondence of the elements of the two sequences. For example, if P = (6)(1,2)(3,5,4) and P' = (4)(2,3)(1,5,6), then Q is the correspondence: $\begin{smallmatrix} 6 & 1 & 2 & 3 & 5 & 4 \\ 4 & 2 & 3 & 1 & 5 & 6 \end{smallmatrix}$ or Q = 2 3 1 6 5 4 = (1,2,3)(4,6)(5). We note that Q is not unique - if we rewrite P' as (4)(3,2)(1,5,6), we obtain Q' = (1,3)(2)(4,6)(5), which is Q·(2,3) corresponding to the fact that 2 and 3 have been interchanged in P'.

Clearly, if a group has enough permutations, then all possible Q are in it and we have a positive answer to our question. It certainly holds in S_n and it is not too hard to see that it holds for the constructible group (i.e. the group of positions of the cube obtained by taking it apart and reassembling it). Of course we must treat the edges and corners separately - a 3-cycle of edges is not conjugate to a 3-cycle of corners.

Uldis Celmins asked if the inverse problem has a positive answer in the cube group. We show the answer is negative. Let P be a position consisting of a 7-cycle of corners and an 11-cycle of edges (e.g. the position α on page 32), say P = (C1,C2,...,C7)(C8)(E1,E2,...,E11)(E12). Let P' be P with two corners exchanged, say P' = (C2,C1,...,C7)... Then P and P' are both achievable positions on the cube and the conjugating Q is (C1,C2) which is an odd permutation and hence not in the group. Now there are other conjugating elements corresponding to rewriting P'. These correspond to multiplying Q by $(C2,C1,...,C7)^i(E1,E2,...,E11)^j$ and hence are still odd. Hence any possible conjugating Q is not in our group and so P and P' are not conjugates.

The above analysis depends on the fact that the conjugating element is determined on each position since P and P' act on almost all the corners and edges. Other examples arise, e.g. if P contains an 8-cycle of corners and P' is P with one corner twisted. However, if P leaves either (two corners and an edge) or (two edges and a corner) fixed, then there is enough freedom of choice for Q that we can make Q be in the group. Thus all 3-cycles of edges are conjugate to one another, etc. Indeed, this point was why we only had to find one 3-cycle of edges.

M. WREATH PRODUCTS.

On page 31, I mentioned that the cube group is an example of a wreath product of groups. Peter M. Neumann has written a chapter on "The Group Theory of the Hungarian Magic Cube" as part of the lecture notes on "Groups and Geometry". (See the Bibliography for details.) Don Taylor's preprint "The Magic Cube" and the Neumann chapter discuss the use of the wreath product and I have finally come to more or less understand it.

Neumann describes the cube group as a subgroup of the constructible group which is the group of positions which can be obtained by taking the cube apart and putting it back together. Then none of the restrictions described on page 12 are applicable and the constructible group has $12 N = 8! \ 12! \ 3^8 \ 2^{12} = 5 \ 19024 \ 03929 \ 38782 \ 72000 = 2^{29} 3^{15} 5^3 7^2 11$ elements. In group theory, we say the cube group has index 12 in the constructible group. (See p 39 for an earlier use of index.) We now describe the constructible group Q. Q is the <u>direct product</u> of its action on corners and its action on edges. That is, the actions are independent (because they act on different sets) so each element of Q can be expressed as a pair consisting of a corner action and an edge action. We denote this by writing $Q = Q_c \times Q_e$.

Recall that S_n is the group of all n! permutations of n objects (p 9) and let Z_n denote the cyclic group of n elements (p 10) - e.g. the group generated by an n-cycle or the group of rotations of a regular n-gon. Let Z_n^k denote the direct product of k examples of Z_n. Its elements are then k-tuples $(a_1, a_2, a_3, \ldots, a_k)$ which combine by independently combining each component separately. <u>Exercise.</u> show that the slice-squared group is the same as Z_2^3. The <u>wreath product</u> Z_3 wr S_8 is a way of defining a multiplication on the product set $Z_3^8 \times S_8$ which corresponds precisely to the way in which the 8 corners can be arbitrarily permuted in position while each corner can be 'independently' changed in orientation. 'Independently' must be in quotes because the multiplication of orientation changes does depend on where the pieces have been moved to by the previous permutation.

Conceptually we have an orientation (i.e. an element of Z_3) attached to each object being permuted - whether a symbol (i.e. piece) or a location. (Recall the dichotomy between permutations acting on symbols and on locations which was discussed in section 4.) We then have 24 oriented objects in the product set $Z_3 \times \{1,2,\ldots,8\}$ and this is the set on which we shall define $Z_3^8 \times S_8$ as acting in the following manner. Consider any element $(a_1, a_2, \ldots, a_8)$ of orientation changes (i.e. elements of Z_3) with any permutation P of S_8. When this is applied to an oriented object (a, i), the result is $(a_{P(i)} \cdot a, P(i))$. So if the object is carried to object j, the j-th orientation change is applied to the orientation at i to get the new orientation at j. It is difficult to explain the behaviour in words, but perhaps the best way to visualise the wreath product is as the U diagrams used on page 42. The arrows show the permutation of pieces being done just as expected and the signs attached to the arrows show that an orientation change takes place as the piece is moved. So the new orientation at j depends on the previous orientation of the piece at i which is coming to j and on the orientation change associated with the movement from i to j. Once the action of $Z_3^8 \times S_8$ on the oriented objects is defined, it is not difficult to verify that one has a group structure and this is the wreath product.

Thus $Q_c = Z_3$ wr S_8 and $Q_e = Z_2$ wr S_{12} and this gives a fairly explicit description of the constructible group. Getting the cube group is just the same argument that we used on pages 8,9,12,17.

The centre of Q is not hard to find now. One must find the centre of Q_c and of Q_e. It is easy to see that because we have the complete permutation group S_8 in $Q_c = Z_3$ wr S_8, then an element of the centre must have all its orientation changes the same and its permutation must be in the centre of S_8. Now PQ = QP for all Q if and only if $Q^{-1}PQ = P$ for all Q, that is to say, every conjugate of P must be just P itself. But our study of conjugation and cycle structure in the previous section shows that the conjugates of P in S_8 are all the permutations with the same cycle structure and there are lots of them. Even restricting to

the cube group does not change this. Thus the only possible central
elements are those which leave everything in place but change all
orientations in the same way. There are 2 of these in Q_e and 3 in Q_e,
but only the former occur in the cube group, and these are the identity
and the 12-flip. (See pp 28, 31, 35, 48.)

N.   THE 2 × 2 × 2 CUBE.

Gerzson Kéri asserts that he can get the pieces correctly positioned
relative to one another in at most 7 moves. Thistlethwaite says that
he can obtain almost any twisting of corners on the 3 × 3 × 3 cube
in at most 18 moves, but he is leaving edges fixed. So we should already
have a method which works in at most 25 moves and this should be be much
improved if we had a repertoire of twists, ignoring edge effects.

Problem. Determine the diameter of the 2 × 2 × 2 cube group.

Because we have no fixed orientation, we can choose the BLD piece
be correct and then only make F, R, U turns. Reasoning as on page 26,
there are at most $1 + 9 + 9 \cdot 6 + 9 \cdot 6^2 + \ldots + 9 \cdot 6^{n-1} = 1 + 9(6^n-1)/5$
positions after n moves. Setting this equal to 36 74160 (see pp 28,29,31),
we get n = 8.11, so some positions will require at least 9 moves.

5.11   MISCELLANY.

Krystyna Dałek asks about the groups generated by commutators of
basic moves. For example, what is <[F,R],[F',R']>? This acts on five
edges and six corners. It is not hard to see that: edges do not change
orientation; all 5!/2 even permutations of edges are achievable; the corners
form two orbits of order 3; the action on one 3-orbit determines the action
on the other; all 3! permutations of an orbit can be achieved without
disturbing edges; all corners of an orbit can be twisted the same way
without affecting edges; all 27 orientations of one orbit can be achieved
but affecting edges. It is not clear whether all the orientations can be
achieved without affecting the edges or not.

Dałek also asks if the commutator subgroup (pp 18,27,28) is generated
by the commutators of basic moves. Recall that the commutator subgroup
is generated by all possible commutators.

Uldis Celmins has given a graphical analysis of the result of
A. Taylor (p 29) regarding other magic polyhedra. He assumes that each
corner has three edges at it, no two faces have more than one edge in
common and each face has at least four edges, and then he gets the result
of Taylor. It is not clear if all these assumptions are necessary or
if BCG's report of Taylor's work left out some of the assumptions.

I have occasionally seen cubes which had some pieces stuck together
from the glue used on the cover pieces (2,3). This leads to interesting(?)
questions as to what patterns are achievable. It may not be possible to
turn some faces either all the time or just sometimes.

Dave Fyfe said that his eight year old son had been so dismayed at
messing up a cube that he tried to restore it by peeling off some of
the coloured stickers. How would you recognize and correct sticker
rearrangement? I have even heard of people doing this to other people's
cubes as a subtle and malicious form of torture.

Frank Barnes observes that the idea of shifting a process described
on page 36 can be used to solve the 'misprint problem'. That is, if a
process has a single error in it, one can examine the rotations of it.
When the error comes last, it is easy to see what it should be corrected to.

Joe Buhler and David Sibley are preparing an article on the
n-dimensional magic cube.

One can get patterns with all outer row and column sums the same
on both sides of the Magic Domino, but only with different values on
the two sides,e.g.

```
1 2 7          9 8 3
6 5 2 opposite 4 5 8
3 6 1          7 4 9 .
```

A STEP-BY-STEP SOLUTION TO RUBIK'S MAGIC CUBE

The notation explained below is the same as given in section 3 of my "Notes on Rubik's Magic Cube". Some familiarity with section 3, with cycles and even permutations (section 4) and with the basic processes P_1, P_2, P_3, P_4 and conjugation (section 5) will help you to understand this method, but you can apply the solution without any of this knowledge.

You may want to practice the individual steps of this method separately before trying to restore the cube completely. It is especially easy to see what is happening if you start with the cube correct, either by taking it apart or by getting (bribing?) someone to do it for you.

The six faces of the cube are labelled Right, Left, Front, Back, Up, Down and are abbreviated to their initials R, L, F, B, U, D. The centres of the faces will remain fixed in space during each substage, so we always know which colour is R - namely the colour of the centre of the R face. Pairs and triples of letters will define both positions and pieces, as in Figure S1. E.g. UF is the edge between the U and F faces, URF is the corner of the U, R and F faces. At the beginning, each piece is in its correct position (or place), but they soon move about. We will say UF is at RB if the UF piece is at the RB position with the U side of UF in the R face.

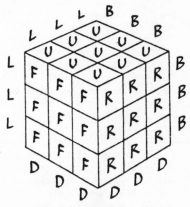

FIGURE S1

The letters R, L, F, B, U, D are also used to describe turns of the faces of the cube. E.g. R is a 90° clockwise turn of the R faces, as shown in Figure S2. Note that DFR is now at FUR, DR is at FR, etc. Each face rotation is taken 90° clockwise as viewed when looking at the face from outside the cube. R^2 will denote a 180° turn (in either direction!) and R' will denote a 90° anticlockwise turn of the R face. E.g. to correct Figure S2, we will want to apply R'. A sequence of moves, such as RUF means apply first R, then U then F. Note that RR = R^2, RRR = R^3 = R', RR' = R'R is the process I of doing nothing. RU is shown in Figure S3. R·U is sometimes used for RU, etc.

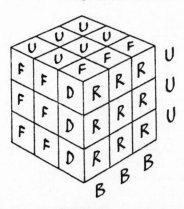

FIGURE S2

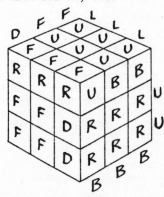

FIGURE S3

THE METHOD

Choose a colour to do first and hold that face centre up. (I often choose white as being the easiest colour to see.)

The first three stages are described for one piece but each must be applied four times with the whole cube turned 90° between applications. The descriptions are for the hardest cases and are memorable rather than efficient. With experience you can find more efficient processes. If your cube has been reassembled incorrectly, The Method may fail at stage 4, 7, or 8. If the coloured stickers have been deranged, The Method may fail at any stage!

1. Put U edges correctly in place.

Let us suppose that we want to put the UF edge piece into its proper place. First find the piece. We shall describe how to get it into place without disturbing any other U edges. But if UF is already somewhere in the U face, we first move it off the U face by turning the side face which contains it. So we have the UF piece not in the U face.

    1.a.   If UF is in the D face with its U side down, then turn D until UF is in the F face and apply F^2. E.g. if UF is at DB, apply D^2F^2.

    1.b.   If UF is in the D face with its F side down, then turn D until UF is in the F face and apply F'U'RU. E.g. if UF is at BD, apply D^2F'U'RU.

    1.c.   If UF is in the middle layer, then it can be rotated into the U face by either of two side turns. One of these side face turns would put the U side of UF in the up direction. First turn the U face so that the UF <u>position</u> is in this side face, then turn the side face and then turn the U face to put the UF piece in place. E.g. if UF is at RB, apply U^2BU^2, while if UF is at BR, apply U'R'U.

You should now have a correct + pattern of U edges.

2. Put U corners correctly in place.

Let us suppose that we want to put the URF corner piece into its proper place. First find the piece. We shall describe how to get it into place without disturbing any U edge or any other U corner. But if URF is already somewhere in the U face, we first move it off the U face by turning a side face that contains it, then applying D^2 and then turning the side face back. (It is easier if you ensure that the U side of URF does not go down.) So we have the URF piece in the D face.

Now turn the D face so that our piece is at the FRD corner.

    <u>2.a.</u>   If URF is at FRD, apply D'R'DR

    <u>2.b.</u>   If URF is at RDF, apply DFD'F'

    <u>2.c.</u>   If URF is at DFR, apply FD^2F'·D^2·R'DR

You should now have a solid U coloured U face with its sides agreeing with the side face centres. Now turn the whole cube over so the correct face is now down (i.e. D). It will remain D through the rest of the method.

3. Put middle layer edges correctly in place.

Let us suppose that we want to put the FR edge piece into its proper place. First find the piece. We shall describe how to get it into place without disturbing the D face or any other middle layer edge. But if FR is already somewhere in the middle layer, we move it into the U face as follows: turn the whole cube so that the offending piece is at the FR position, apply $B'U·R^2U^2R^2U^2R^2U^2·U'B$ (which moves the piece to BU), then turn the whole cube to where it was. So we have the FR piece in the U face.

    3.a.   If the F side of the FR piece is up, turn the U face until the FR piece is at UL, then apply $LU·U^2F^2U^2F^2U^2F^2·U'L'$.

    3.b.   If the F side of the FR piece is not up, turn the U face until the FR piece is at BU, then apply $B'U·R^2U^2R^2U^2R^2U^2·U'B$.

You should now have the bottom and middle layers correct.

4. Orient U edges.

An even number of U sides of U edges will now be up. By turning the cube, you can apply one of the following.

    4.a.   If the pieces at UB and UF are incorrectly oriented, apply B·LUL'U'·B'.

    4.b.   If the pieces at UB and UL are incorrectly oriented, apply B·ULU'L'·B'.

    4.c.   If all four pieces are incorrectly oriented, apply 4.a, turn the cube 180° and then apply 4.b.

You should now have a + pattern of U edge faces in the U face.

5. Make the U edges into an even permutation.

Examine the 4 U edge pieces and compare their side faces with the side face centres. Turn U so that UF is at UF.

If a total of two U edge pieces are now in their proper places, apply U.

You should now have 0, 1 or 4 U edge pieces in place.

6. Put U edges in place.

    6.a.   In one case out of six, the U edges are now all in place. Do nothing.

    6.b.   In four cases out of six, there will be just one U edge correctly in place and the other three want to be cycled. Turn the cube so the correct piece is at UL.

    6.b.i.   If the three U edges want to be cycled clockwise, as in Figure 84, apply $R^2D'·U^2R'LF^2RL'·DR^2$.

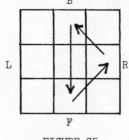

FIGURE S4   FIGURE S5

    6.b.ii.   If the three U edges want to be cycled anticlockwise, as in Figure 85, apply $R^2D'·R'LF^2RL'U^2·DR^2$.

    6.c.   In one case out of six, there will be no U edge pieces correctly in place and two adjacent pairs of edges want to be exchanged. Turn the cube so you want to exchange UF with UR and UL with UB, as in Figure S6. Apply $R^2D^2B^2D·L^2F^2L^2F^2L^2F^2·D'B^2D^2R^2$.

You should now have all the U edges correctly in place.

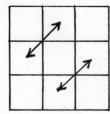

FIGURE S6

7. Put U corners in place.

    7.a.   In one case out of twelve, the U corners are now all in their correct places, though perhaps disoriented. Do nothing.

    7.b.   In eight cases out of twelve, one U corner is in its correct place, though possibly disoriented, and the other three want to be cycled. Turn the cube so the correctly placed piece is at the URF corner.

    7.b.i.   If the three U corners want to be cycled clockwise, as in Figure S7, apply L'·URU'R'·L·RUR'U'.

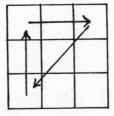
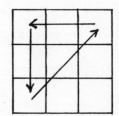

FIGURE S7   FIGURE S8

7.b.ii.   If the three U corners want to be cycled anticlockwise, as in Figure S8, apply URU'R'·L'·RUR'U'·L.

7.c.   In two cases out of twelve, there will be no U corners correctly placed, and there will be two pairs of adjacent corners that want to be exchanged. Turn the cube so the exchanges are along the UF and UB edges of the cube, as in Figure S9. Apply B·LUL'U'·LUL'U'·LUL'U'·B'.

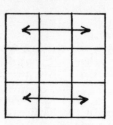

FIGURE S9

7.d.   In one case out of twelve, there will be no U corners correctly placed and the two diagonal pairs of corners want to be exchanged, as in Figure S10. Apply R'B²·FRF'R'·FRF'R'·FRF'R'·B²R.

You should now have all the U corners in their correct places, though possibly disoriented.

<u>8.   Orient U corners.</u>

Turn the cube so some incorrectly oriented U corner piece is at the URF position.

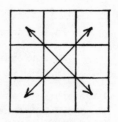

FIGURE S10

8.a.   If the piece at URF wants to be twisted clockwise (viewed from outside the cube), apply FDF'D'·FDF'D'.

8.b.   If the piece at URF wants to be twisted anticlockwise. apply DFD'F'·DFD'F'.

This should make the piece at URF be correctly oriented, though it temporarily confuses the rest of the cube.

Now turn <u>just the U face</u> to bring another incorrectly oriented U corner piece into the URF position. Apply 8.a or 8.b as needed. Continue in this way until all U corner pieces are correctly oriented. You will then need to turn the U face to bring UF back to the UF position and the cube should now be completely correct!

<u>9.   (Optional step).</u>   Scream HOORAY!   Buy a round of drinks. Send me a cheque. Tell the orderlies that they can let you out now. Etc., etc.

* * *

BIBLIOGRAPHY

    I list here all items I know of. Many of these are working papers,
drafts or preprints, so I have given the address of the author or an author
or such information as I know. References in the text to persons not
given here indicate 'personal communication'.

Angevine, James. Solution for the Magic Cube Puzzle. 11 pp, 1979?; 5 pp,
    1980? Available for $2.00 and SAE from Logical Games (address on p 37)
    or The Angevine Solution, PO Box 1356, Manassas, Virginia, 22110, USA.
Beasley, J. D. (7 St James Rd, Harpenden, Herts, UK). A solution to the
    twisted cubes. 12pp, 1980?
Bedecs, Éva. A kocka kockázata. Magyarország, year 16, week 47, number 826
    (25 Nov 1979) 32.
BCS: Benson, David, John H. Conway & David Seal (all at DPM&MS, 16 Mill
    Lane, Cambridge, CB2 1SB, UK). Solving the Hungarian Cube. 8pp,
    1980?, preprint intended for the Journal of Recreational Mathematics?
BCG: Berlekamp, Elwyn R., John H. Conway & Richard K. Guy (Guy is at
    Mathematics, Univ of Calgary, 2920 24th Ave NW, Calgary, Alberta,
    T2N 1N4, Canada). Winning Ways. This said to be at the publishers
    and may appear in 1980. Harcourt, Brace? Section on: The Hungarian
    Cube - Bűvös Kocka. First draft, 16pp, 1979; Second draft, 15pp, 1980.
Bremer, Hanke. Ein paar Würfeldrehungen. 5pp, Aug 1978. (Sent to me by
    Roberto Minio, Springer-Verlag, Postfach 10 52 80, D-6900 Heidelberg 1,
    West Germany.)
Buhler, Joe (Mathematics, Penn. State Univ., University Park,
    Pennsylvania, 16802, USA). So you want to solve the Magic Cube?
    4 pp, 1980?
---------- & Sibley, David. (Article on the n-dimensional problem).
Cairns, Colin & Dave Griffiths (ICL Dataskil, 118-128 London St, Reading,
    RG1 4SU, UK). Teach yourself cube-bashing. 6pp, Sep 1979. (An
    extended version of the material in Truran's second article.)
Celmins, Uldis (Combinatorics, Univ of Waterloo, Waterloo, Ontario, N2L 3G1,
    Canada). The Hungarian Magic Cube Puzzle. Preprint, 11pp, 1979.
    To appear in: Proceedings Joint Canada-France Combinatorial Conference
    to be published by Discrete Mathematics.
Conder, Marston D. E. (Mathematical Institute, 24-29 St Giles, Oxford,
    OX1 3LB, UK). On generating the group of the 'Magic Cube' by two
    elements. Revised version, 4 pp, 1980.
Dauphin, Michel (Lycée de garçons de Luxembourg). Un cube pas comme les
    autres. Mathématique et Pédagogie, No 24 (Nov-Dec 1979) 23-29.
De Koven, Bernie. The Magic Cube. Games 3:6 (No 14)(Nov-Dec 1979) 78.
Gaskin, John (5 Meadow Lane, Fetcham, Leatherhead, Surrey, UK). Cubist
    Rescue Service - Emergency Kit, "Getting back to Square One".
    4pp, 1979.
Howlett, G. S. Pentangle - "Magic Cube", A guide to the solution. 2pp.
    Available from Pentangle (address on p 22).
Ideal Toy Co. Ltd. (Fishponds Road, Wokingham, Berks, RG11 2QR, UK).
    Rubiks Cube, p 20 of their 1980 catalogue.
Ishige, Terutoshi. Japanese Patent 55-3956 (for the 3 × 3 × 3). 6pp
    numbered 183-189, dated 29 Mar 1977; 21 Oct 1978; 28 Jan 1980.
------------------. ---------------- 55-8192 (for the 3 × 3 × 3). 3pp
    numbered 193-195, dated 12 Oct 1976; 26 Apr 1978; 3 Mar 1980.
------------------. ---------------- 55-8193 (for the 2 × 2 × 2) 4pp
    numbered 197-200, dated 12 Mar 1977, 4 Oct 1978, 3 Mar 1980.
Jackson, 3-D (Bradley W.) (Mathematics, Univ. of California, Santa Cruz,
    California, 95064, USA). The Cube Dictionary. 15 pp. 1979.

Kéri, Gerzson (MTA SZTAKI, H-1502 Budapest, Kende u 13-17, Hungary). A bűvös
    kocka matematikája-I. Középiskolai Matematikai Lapok 60 (1980) 97-107.
----   ---------- - II. Ibid 60 (1980) 193-198.
----   A Bűvös Kocka Problemakörröl. Working Paper, 39pp, Nov 1979.
Maddison, Richard (Mathematics, Open University, Milton Keynes, MK7 6AA,
    UK). The Magic Cube. 2pp, Nov 1979.
Nelson, Roy (same address as Maddison). The Hungarian Cube. 9pp, Nov 1979.
Neumann, Peter M. The Group Theory of the Hungarian Magic Cube. Chapter
    19, pp 299-307 of: Peter M. Neumann, Gabrielle A. Stoy & E. C. Thompson,
    "Groups and Geometry". Lecture Notes, April 1980, available from the
    address given for Conder at £2.00 plus £1.22 for UK postage or £0.70
    for overseas postage.
Ollerenshaw, Dame Kathleen. The Hungarian Magic Cube. Bull Inst Math Appl
    16 (Apr 1980) 86-92.
Perjés, Zoltán (Central Research Institute for Physics, Budapest 114,
    PO Box 49, Hungary). On Rubik's Cube. Revised version, 19pp, Jan 1980.
Rubik, Ernö. Hungarian Patent 170062. 8pp, dated 30 Jan 1975; 28 Oct 1976;
    31 Dec 1977.
Sakane, Itsuo. (Article in Japanese). Asahi Shimbun (1 June 1980) 15.
Singmaster, David. The "Magic Cube". 2pp, 1979. (Information sheet also
    containing the original Hungarian advertising and a translation.)
--------. Six-sided Magic. The Observer (17 June 1979) 40.
    (Follow-up note on 27 Jan 1980).
--------. The Hungarian Magic Cube. Mathematical Intelligencer
    2:1 (1979) 29-30 (& cover).
--------. The Magic Cube. Games & Puzzles, No 76 (Spring 1980) 4-6.
Taylor, Donald E. (Pure Mathematics, Univ of Sydney, New South Wales, 2006,
    Australia) The Group of a Coloured Cube. 8pp, 1978.
------ The Magic Cube. Revision of above, 18pp, Nov 1978.
------ Rubik's Cube. 18pp, Apr 1980.
------ Secrets of the Rubik Cube - A Guided Tour of a Permutation
    Group. To appear.
Truran, Trevor. Cunning of the Eastern Bloc(k). Computer Talk (5 Sep 1979) 8
-----. Solving that Eastern Bloc(k). Ibid (7 Nov 1979) (An outline
    of Cairns & Griffiths).
-----). Two letters in response to the second article from G. Blow
    and M. S. Maxted. Ibid (9 Jan 1980) 9.

ADDITIONAL BIBLIOGRAPHY

Adams, Arthur. (125 Hall Lane, Upminster, Essex, RM14 1AU, UK). The
    Hungarian Magic Cube. 21 pp, Sep 1980.
(Anonymous). The magic cube. New Hungarian Exporter (Periodical of the
    Hungarian Chamber of Commerce) 30:2 (Feb 1980) 18 (+ photo on the back
    cover).
(Anonymous). Mathematische Kabinett: Wir enträtseln den Zauberwürfel. Bild
    der Wissenschaft, Year 17, No. 11 (Nov 1980) 174-177.
--------. ---------------------: Verflixt - nochmal. Ibid., Year 17, No.
    12 (Dec 1980) 180-187. Another article to appear in April 1981.
(Anonymous). 30 "bons" jeux au banc d"essai. Jeux & Stratégie, No. 6 (Dec
    1980/Jan 1981) Photos on 40-41; Le domino hongrois, 44; Rubik's cube,
    44-45; Ad. 59.
(Anonymous). Cube Solution. 1 p. (Available from Marith, 2 The Waterloo,
    Circencester Glos., UK).
(Anonymous). Erfreue dich der Symmetrie. Der Spiegel, Year 35, No. 4 (19
    Jan 1981) 181-184.
Blewett, William J. (12162 Wilsey Way, Poway, California, 92064, USA).
    Solving Rubik's Cube. 18 pp, 1980.
Buhler, Joe. (Mathematics, Penn. State Univ.,University Park, Pennsylvania,
    16802, USA). So you want to solve the Magic Cube? 4 pp. (Amendment of
    entry on B1 of my Notes.)

Claxton, Mike. Solution to the Magic Cube. 8 pp, 1979.

Craats, Jan van de. Magische kubus: voor wiskundige maniakken. NRC Handelsblad (Amsterdam) (1 Oct 1980) 11.

Deledicq A. & J.-B. Touchard. Le Cube Hongrois - Mode d'Emploi. IREM, Paris VII, T. 56/55, 3 eme Etage, 2 Place Jussieu, F-75005, Paris, France. 74 pp, 4th printing, Dec 1980.

Endl, Kurt. Rubik's Cube (in German). Würfel-Verlag GmbH (Postfach 6627, D-6300 Giessen, Germany). 48 pp. Sep 1980.

Engelhardt, Matthias. (Institute fur numerische Mathematik, Roxeler Strasse 64, D-4400 Munster, Germany). My way to twist Rubik's Cube. 8 pp,1980.

Golomb, Solomon W. Rubik's cube and a model of quark confinement. 8 pp,1980.

Halberstadt, Emmanuel. Cube hongrois et théorie des groupes. Pour la Science, No. 34 (Aug 1980) 23-36.

Hofstadter, Douglas. Metamagical Themas. Scientific American (Mar 1981) 20-39.

Jullien, Pierre. Le cube hongrois. Education et Informatique, No. 4 (unknown date) 21-23.

Juppenlatz, Peter. Mein Hit, der hat acht Ecken. Stern, Year 34, No. 1 (23 Dec 1980) 140-143.

Kéri, Gerzson. A búvös kocka matematikája - II. Középiskolai Matematikai Lapok 60 (1980) 193-198. (Part I was similarly titled. I had previously seen the proofs which gave the title as on 66 of my Notes.)

Last, Bridget. A Simple Approach to the Magic Cube. Tarquin Publications (Stradbroke, Diss, Norfolk, UK) 20 pp, 1980. ₤0.50 (+ ₤0.90 post & packing per order).

(Minio, Roberto.) Cube News. The Mathematical Intelligencer. 2:4 (1980) 161-162.

Morris, Scot. Games: Ernö Rubik's Magic Cube. Omni, 2:12 (Sep 1980) 128-129.

----------------: Inside Rubik's Cube. Omni, 3:1 (Oct 1980) 193.

----------------: Games column. Rubik's Update. Omni 3:5 (Feb 1981) 129.

Naumann, Swantje. Puzzle um acht Ecken. Die Zeit (12 Dec 1980) 58.

Pracontal, Michel de. Un cube fou, fou, fou. Science et Vie, No. 753 (Jun 1980) 135-139.

Ristanovic, Dejan. Magicna kocka na domaci nacin. Galaksija (Dec 1980). (This is a popular science magazine from Belgrade.)

Roddewig, Ulrich. Solution Guide for the Hungarian Magic Cube. 9 pp, June 1979, (Only address given is Köln.)

(The Roddewig, Marith & Claxton items were forwarded by Pentangle, who presumably have the addresses.)

Rowley, Chris. The group of the Hungarian Magic Cube. To appear in Proc. First West Australian Algebra Conference. (1981 ?) Preprint of the 12 pp available from the author, Open University London Office, 527 Finchley Road, London, NW3.

Rubinstein, Steve. Puzzle with a twist. San Francisco Chronicle (10 Jan 1981)5.

Sakane, Itsuo. (in Japanese). Asahi Shimbun (18 Jan 1981) 15 and (25 Jan 81).

Taylor, Don. Mastering Rubik's Cube. Book Marketing Australia P/L (195 Bridge Road, Richmond, Victoria 3121, Australia). 31 pp, 1980. 1.95 $Australian (₤1.00.≃ $2.40).

Thistlethwaite, Morwen. The 45-52 move strategy. 4 pp typescript text + 7 pp computer output reduced to 4 pp, July 1980.

----------------. Some remarks on cube subgroups. 3 pp typescript, Aug 1980.

----------------. The Magic Cube. 4 pp typescript, Aug 1980.

----------------. Instructions for restoring cube. 2 pp typescript, Aug 1980. (His address is now: Mathematics - Wandsworth Road, Polytechnic of the South Bank. London. SE1 0AA.)

Thompson, John G. (same address as BCS). Rational functions associated to presentations of finite groups. 10 pp, Sep 1980 (preprint).

Toth, Viktor. A búvös kocka egy gyors redezése. Középiskolai Mathematikai Lapok 60 (1980) 198-200.

Tricot, Jean. Manifestations de groupe. Jeux & Stratégie, No. 6 (Dec 1980/ Jan 1981) 36-38.

Truran, Trevor. A whole new cubist movement. Computer Talk (29 Oct 1980) 8.

Warshofsky, Fred. Rubik's cube - Madness for millions (draft title).
    Reader's Digest. To appear in various editions this spring.

(Weisman, Robert.) Rubik's Cube Puzzle: An Introduction and Solution. Ideal
        Toy Corporation, (184-10 Jamaica Avenue, Hollis, New York, 11423, USA).
        14 pp, 1980.

Zalgaller, V. & C. Vengerskii sarnirnyi kubik (Hungarian pivotted cube).
        Matematiceskii Kruzok (Mathematical Circle) Kvant (volume ?) 2, No. 12
        (1980) 17-21. (With a supplement by V. Dubrovskii.)

Ziemke, Peter-Michael. (Am Hohen Rod 7, D-3500 Kassel, Germany).   Drehbuch
        für (the Magic Cube). 66 pp, Sep 1980.

INDEX

Since the text has grown by accretion and is very dense with material, I have made lots of cross references. Nonetheless I think an index is needed as well.

For references to general classes of processes, e.g two 2-cycles, consult section 5.8 (pp 43-46). For references to particular groups and pretty patterns, look under Groups and Pretty Patterns. There are entries for Problems (and their answers and continuations) and for Figures.

Many concepts are used before they are defined, so the first page reference may not be a definition of the term.

Notation and numeric information precede the alphabetic listing.

The Index does not cover the Introduction nor the Step by Step Solution.